The Meaning of Quantum Gravity

Fundamental Theories of Physics

An International Book Series on The Fundamental Theories of Physics: Their Clarification, Development and Application

The Meaning of Quantum Gravity

by

H.-H. von Borzeszkowski

and

H.-J. Treder

Einstein-Laboratorium für Theoretische Physik
der Akademie der Wissenschaften der D.D.R.,
Potsdam-Babelsberg, D.D.R.

D. Reidel Publishing Company

A MEMBER OF THE KLUWER ACADEMIC PUBLISHERS GROUP

Dordrecht / Boston / Lancaster / Tokyo

Library of Congress Cataloging in Publication Data

CIP

Borzeszkowski, H.-H. von (Horst-Heino von)
The meaning of quantum gravity.

(Fundamental theories of physics)
Bibliography: p.
Includes index.
1. Quantum gravity. I. Treder, Hans Jürgen.
II. Title. III. Series.
QC178.B635 1987 530.1 87–27501
ISBN-13: 978-94-010-8229-7 e-ISBN-13: 978-94-009-3893-9
DOI: 10.1007/978-94-009-3893-9

Published by D. Reidel Publishing Company,
P.O. Box 17, 3300 AA Dordrecht, Holland.

Sold and distributed in the U.S.A. and Canada
by Kluwer Academic Publishers,
101 Philip Drive, Norwell, MA 02061, U.S.A.

In all other countries, sold and distributed
by Kluwer Academic Publishers Group,
P.O. Box 322, 3300 AH Dordrecht, Holland.

Softcover reprint of the hardcover 1st edition 1988

Table of Contents

Preface

In discussing the question of whether General Relativity Theory really needs to be quantized, a simply negative answer cannot be accepted, of course. Such an answer is not satisfying because, first, Einstein's gravitational equations connect gravity and non-gravitational matter and because, second, it can be taken for granted that non-gravitational matter has an atomic or quantum structure such that its energy-momentum tensor standing on the right-hand side of Einstein's equations is formed out of quantum operators. These two facts make it impossible to read the left-hand side of Einstein's equations as an ordinary classical function. This does not necessarily mean, however, that we must draw the conclusion that General Relativity Theory, similar to electrodynamics, could or should be quantized in a rigorous manner and that this quantization has similar consequences to quantum electrodynamics.

In other words, when for reasons of consistency quantization is tried, then one has to ask whether and where the quantization procedure has a physical meaning, i.e., whether there exist measurable effects of quantum gravity.

In accordance with these questions, we are mainly dealing with the discussion of the principles of quantized General Relativity Theory and with the estimation of quantum effects including the question of their measurability. This analysis proves that it is impossible to distinguish between classical and quantum General Relativity Theory for the extreme case of Planck's orders of magnitude. In other words, there does not exist a physically meaningful rigorous quantization conception for Einstein's theory.

Because the quantized Einstein theory for free gravitational fields contains all universal constants known today, namely h, c, and G, it satisfies the Einstein—Heisenberg demands on a unitary theory. A unitary theory should, therefore, basically imply General Relativity. If this is accepted then, as a consequence of the quantum gravity analysis presented here, one has to conclude that a super-GUT (i.e., a GUT including gravity) should not differentiate between quantum and classical physics.

August, 1987 H.-H. von Borzeszkowski
H.-J. Treder

Chapter 1

Quantum Theory and Gravitation

Einstein's General Relativity Theory (GRT) was originally formulated as a classical field theory, and the question if it can or even must be quantized is one of the fundamental questions which has not been finally answered until now. The consideration of this problem is not so much justified by formal analogies between electromagnetic and gravitational theories, and even less by metaphysical belief in the quantrum structure of nature, but by the fact that matter fields must be quantized. This means that the character of the coupling between quantized matter and gravity must be clarified. The main aspect of the problem is whether this requires gravity to be quantized too. If one refers to Einstein's GRT, the problem is now to understand Einstein's equations

$$G_{\mu\nu} \equiv R_{\mu\nu} - \tfrac{1}{2} g_{\mu\nu} R = -\kappa T_{\mu\nu} \tag{1.1}$$

(where $\kappa = 8\pi G/c^4$) in the case of quantized matter, i.e., in the case where the energy—momentum tensor $T_{\mu\nu}$ describes quantized matter.

In a discussion with P. A. M. Dirac* following his lecture (1967), one of us (H.-J. Treder) asked him if it was possible to interpret Einstein's general relativistic equations (1.1) so that the right-hand side of (1.1) represents the expectation value $\langle \mathbf{T}_{\mu\nu} \rangle$ of the tensor $T_{\mu\nu}$ describing quantized matter, while the left-hand side of (1.1) is considered as the classical Einstein tensor

$$G_{\mu\nu} = R_{\mu\nu} - \tfrac{1}{2} g_{\mu\nu} R \tag{1.2}$$

of non-quantized gravitational fields, such that one has to read (1.1) as

$$G_{\mu\nu} = -\kappa \langle \mathbf{T}_{\mu\nu} \rangle \tag{1.3}$$

where $\mathbf{T}_{\mu\nu}$ is the energy—momentum density operator. Dirac rejected this proposal categorically with the remark that the right-hand side of (1.3) would then depend on the choice of the state vector in the Hilbert space (i.e., it would not be Hilbert covariant), while the left-hand side of (1.3) would be a

* Cf., *Fluides et Champ Gravitationnel en Relativité Générale*, Paris, Collège de France, 19—23 juin 1967, pp. 23 etc.

Hilbert-space scalar. Therefore, any interpretation of (1.1) in the sense of (1.3) and even any use of (1.3) in the sense of a semiclassical approximation must be excluded for mathematical reasons. If (1.3) were to be used, paradoxical consequences would result.

One has therefore to follow an alternative method. Either one has to quantize gravitational fields, i.e., to read Einstein's equations (1.1) as operator equations

$$\mathbf{G}_{\mu\nu} = -\kappa \mathbf{T}_{\mu\nu} \tag{1.4}$$

(this implies, of course, the existence of 'gravitons' as zero rest mass field quanta) or one must relate the expectation values of the matter tensor, $\langle \mathbf{T}_{\mu\nu} \rangle$, to the average values of the classical Einstein tensor, $\langle G_{\mu\nu} \rangle$,

$$\langle G_{\mu\nu} \rangle = -\kappa \langle \mathbf{T}_{\mu\nu} \rangle. \tag{1.5}$$

If this viewpoint is accepted, quantum investigations have to decide which of these possibilities describes the coupling of matter to gravity; or else to show their physical equivalence. In doing so, one could determine whether the quantized GRT given by (1.4) is really (i.e., in its physical content) different from the non-quantized GRT given by (1.5). The term 'in its physical content' is used to emphasize that we do not pose a purely mathematical question, but ask whether there follow *experimentally testable* effects from quantized GRT which do not result from classical theory.

For example, in electrodynamics, the operator equations

$$\mathbf{F}^{\mu\nu}_{\ ,\nu} = \mathbf{j}^{\mu}, \tag{1.6}$$

where

$$\mathbf{j}^{\mu} = e(\bar{\boldsymbol{\psi}} \gamma^{\mu} \boldsymbol{\psi}) \tag{1.7}$$

is Dirac's operator of current density and $\mathbf{F}^{\mu\nu}$ the field strength operator of the Maxwell field, differ physically from the mathematically weaker equations

$$\langle F^{\mu\nu}_{\ ,\nu} \rangle = \langle \mathbf{j}^{\mu} \rangle. \tag{1.8}$$

It should be mentioned that Planck, in his lectures 'Vorlesungen über Elektrodynamik', stressed that Maxwell's field equations cannot be derived uniquely from Faraday's integral form of the basic laws of electrodynamics (i.e., from the Faraday law of induction and the Ampére law). Maxwell's equations contain much more information and hypotheses than Faraday's and Ampére's integral laws, implying, for instance, Maxwell's stress tensor and the Poynting vector of the momentum current.

Quantum electrodynamics, i.e., electrodynamics which implies Einstein's photon hypothesis, requires the validity of the strong equations (1.6). In this case, the matter field (Dirac's spinor field) and the Maxwell field are quantized. Accordingly, there exist electrons *and* photons, and equations (1.6) imply energy—momentum conservation in individual elementary processes.

Version (1.8) contains, in contrast to (1.6), only a statement on average values, i.e., on a statistical ensemble of elementary particles and processes. This concept of quantum electrodynamics was proposed by Bohr *et al.* (1924). It attempts to couple atomistic matter to nonquantized electromagnetical fields. It results in pure probability laws in the sense of classical statistics. In this approach, energy—momentum conservation does not hold true for individual processes, but should hold statistically, as an average over many such processes. Due to the Compton and Bothe—Geiger experiments showing the validity of the energy—momentum conservation in individual processes, the Bohr—Kramers—Slater approach was abandoned. The following development led to quantum electrodynamics in the sense of operator equations (1.6).

The gravitational equation (1.3) implies, of course, the same consequences for the energy-momentum laws as in the Bohr—Kramers—Slater proposal. Our thesis, which we will justify in the following chapters, is that by virtue of the Planck—Rosenfeld uncertainty relation between the measurable values of the gravitational field $g_{\mu\nu}$ and the linear distances L_0 over which the field measurement is performed,

$$\Delta g L_0^2 \gtrsim \frac{G\hbar}{c^3}, \tag{1.9}$$

the exact validity of the energy—momentum conservation cannot be tested experimentally in individual processes. More generally, all measurable consequences of GRT are described by the average equations

$$\langle G_{\mu\nu}\rangle = -\kappa\langle \mathbf{T}_{\mu\nu}\rangle \tag{1.10}$$

in the same manner as by the equations

$$\langle \mathbf{G}_{\mu\nu}\rangle = -\kappa\langle \mathbf{T}_{\mu\nu}\rangle \tag{1.11}$$

following from the operator equations. There is accordingly no physical difference between nonquantized and quantized GRT.

To verify this thesis we shall, taking the Bohr—Rosenfeld standpoint on measurement (Bohr and Rosenfeld, 1933), analyse the measurement process in quantum gravity and compare it to some predictions of quantum gravity calculations. As far as measurement is concerned, our point of view is epistemologically opposite to Wigner's viewpoint (cf. Wigner, 1982).

Our thesis is that if and only if the strong principle of equivalence is satisfied, there is no difference between quantized and non-quantized gravitational theory. This also means of course that for each theory using more than ten $g_{\mu\nu}$ to describe gravity, i.e., for each theory violating the strong equivalence of gravity and inertia, there exists a physical difference between its quantized and non-quantized versions.* Moreover, our thesis does not necessarily exclude 'global quantum effects'. Such effects result, however, from additional background information, as boundary conditions formulated, e.g., by Sommerfeld's *Ausstrahlungsbedingung*. This coincides with the well known fact that there is a cosmological difference between Einstein's GRT and Einstein—Rosen bimetric interpretations of GRT or Rosen's bimetric generalizations of GRT.

It must also be stressed that in a unified physical theory describing the coupling of matter and gravity in the sense of Faraday, Riemann, Einstein and Heisenberg so that the dualism 'gravity and matter' is cancelled, there can also be a difference between quantized and nonquantized 'gravity'. (For a discussion of such generalized conceptions of gravity, see: Bergmann (1959, 1979), Ivanenko (1979a, b), Wheeler (1966, 1968).) In general, such a theory again violates the strong equivalence principle. This violation should, however, show up in a modification of the Newtonian gravitational law (Steenbeck and Treder, 1984).

To make our point quite clear, let us make some further introductory comments on the physical content of GRT. Using Einstein's, Fock's, Laudau's, or Møller's affine tensor of the energy—momentum of gravitational fields, Einstein's vacuum equations can be written as

$$W[g_{\mu\nu}] = -\tfrac{1}{2} \frac{1}{\sqrt{-g}} (t_{\mu\nu} + t_{\nu\mu}) \tag{1.12}$$

where $W[g_{\mu\nu}]$ is a quasilinear wave operator defined with respect to the metric $g_{\mu\nu}$ itself. In the second-limit approximation (i.e., also in the high-frequency approximation used later) Einstein's vacuum equations therefore read

$$\sqrt{-g}\Box g_{\mu\nu} = -\tfrac{1}{2}(t_{\mu\nu} + t_{\nu\mu}). \tag{1.13}$$

(We use the harmonic coordinate condition $g^{\mu\nu}_{,\nu} = 0$; $\Box$ denotes the d'Alembert wave operator.) By virtue of the fact that the right-hand side of (1.13) represents gravitational self-interaction, there arises a scattering of gravitational waves by gravitational waves and similar effects. The main point we want to stress here is that, due to nonlinearity of GRT, such effects already occur on the level of the classical (non-quantized) theory.

In electrodynamics, such effects generally arise as a consequence of the quantization of Maxwell's vacuum equations or of the coupled Maxwell—Dirac equations. Photon—photon interaction arising in quantum electrodynamics can however be modelled in classical electrodynamics (not using the photon and electron concepts) by nonlinear modifications which were considered by Schrödinger and by Born and Infeld. This was demonstrated

* For a discussion of the connection existing between generalizations of GRT and the quantum problem, see Mercier *et al.* (1978).

by Heisenberg, Euler, and Kockel (1935, 1936). Under this point of view, a decisive difference between Maxwell's electrodynamics and Einstein's gravitodynamics consists in the fact that the missing nonlinearity of Maxwell's equations is compensated by quantization, while Einstein's equations are already nonlinear in their classical version.

To avoid misunderstandings, it should be mentioned here that in the case of the coupled Maxwell—Dirac theory the underlying field equations are also nonlinear. Indeed, in this case one has to consider the equations

$$(-i\gamma_\mu \partial^\mu - m_0)\psi = e\gamma_\mu \psi(x) A^\mu(x), \tag{1.14}$$

$$F^{\mu\nu}_{\ \ ,\nu}(x) = j^\mu(x) = e\bar{\psi}(x)\gamma^\mu \psi(x). \tag{1.15}$$

For $A^\mu_{\ ,\mu} = 0$ one can deduce from them the following nonlinear integro-differential equations

$$(-i\gamma^\mu \partial_\mu - m_0 - \gamma^\mu \lambda A^{\text{ext}}_\mu)\psi(x)$$

$$= e^2 \gamma^\mu \psi(x) \int \mathrm{d}y D(x-y)\bar{\psi}(y)\gamma_\mu \psi(y). \tag{1.16}$$

Here λ is a parameter of the external field A^{ext}_μ and mass m_0 is renormalized by the requirement that, for $\lambda = 0$, the free equations

$$(-i\gamma^\mu \partial_\mu - m_R)\psi(x) = 0 \tag{1.17}$$

must be satisfied.

Due to their nonlinearity, those classical equations provide the Thomson effect of light scattering by electric charges (electrons). Assuming, however, in accordance with the Bohr—Kramers—Slater proposal that one should not change equations (1.16) to their quantized (i.e., operator) form, then one would obtain neither the Klein—Nishina formula for individual elementary Compton scattering effects nor vacuum quantum effects such as the Euler scattering of light by light. The latter effects are missing due to the linearity of classical vacuum electrodynamics. In quantum electrodynamics one has, in contrast to classical theory, virtual Dirac currents (virtual electron twin pairs) such that the theory becomes nonlinear. This nonlinearity can be modelled macroscopically by the Heisenberg—Euler—Kockel development. It is, of course, not a complete model; i.e., the development does not describe individual elementary processes but again only the average over many processes.

Returning to gravitodynamics, one can state that Einstein's GRT should not require quantization to provide all nonlinear effects, including nonlinear vacuum effects. It is without quantization 'sufficiently nonlinear'. If we nevertheless think of quantization, then this should be done to arrive at a self-consistent coupling of quantized matter and gravity mentioned as an introduction. But this 'consistency procedure' is not unlimitedly possible. In particular, high-frequency considerations, proposed by von Borzeszkowski (1982, 1984), show that one arrives at a region, where quantization of nonlinear GRT loses its physical sense.

As far as pure vacuum GRT is concerned, in the second-limit approximation, where

$$W[\mathbf{g}_{\mu\nu}] = -\tfrac{1}{2}(\mathbf{t}_{\mu\nu} + \mathbf{t}_{\nu\mu}) \tag{1.18}$$

the quantization of the energy—momentum affine tensor

$$\frac{1}{\sqrt{-g}}\mathbf{t}_{\mu\nu} = t_{\mu\nu} \tag{1.19}$$

is equivalent to the quantization of the gravitational field $g_{\mu\nu}$ itself. Our statement that pure vacuum GRT needs no quantization means accordingly that the quantized vacuum equations (1.6), or better, the equations

$$\langle W[\mathbf{g}_{\mu\nu}]\rangle = -\tfrac{1}{2}\langle(\mathbf{t}_{\mu\nu} + \mathbf{t}_{\nu\mu})\rangle \tag{1.20}$$

resulting from (1.6) should not be distinguishable from the non-quantized equations

$$\langle W[g_{\mu\nu}]\rangle = -\tfrac{1}{2}\langle(t_{\mu\nu} + t_{\nu\mu})\rangle. \tag{1.21}$$

To our minds, the position briefly outlined above is also corroborated by investigations which start with another physical premise. Indeed, starting from the Einstein—Hilbert action

$$I = \int \sqrt{-g}\, R\, \mathrm{d}^4x, \tag{1.22}$$

which corresponds to the field equations (1.1), and investigating whether one is led to a renormalizable and unitary quantum theory, it was shown that the resulting quantum theory is not renormalizable. This raises the question as to whether this result is due to a technical defect of the quantum procedure used or if it speaks for the conjecture that gravity is essentially classical. In the latter case one should not look for rigorous quantum methods which remove this defect but only for methods guaranteeing that such quantum procedures work in certain approximations where we know that quantum matter in a curved spacetime is physically reasonable, and where we have therefore to perform a self-consistent coupling to gravity. However, independently of one's viewpoint, one has to answer this question by attacking this problem of renormalizability; and a lot of technical work was and is being done to overcome it.

One way which might lead us out of this dilemma is to look for modifications of the action integral (1.22). The most promising idea, discussed intensively in recent years, is to consider higher-derivative modifications of the form $\propto R^2$,

$$I = \int \mathrm{d}^4x \sqrt{-g}\,[R + l^2(\alpha R^2 + \beta R_{\mu\nu} R^{\mu\nu})], \tag{1.23}$$

where l is a length parameter. Such modifications make the theory perturbatively renormalizable. Unfortunately, now the perturbative unitarity is lost because the classical theory given by (1.23) allows small fluctuations with negative energy. The introduction of an indefinite metric on the space of states avoids states of negative energy in the quantum theory, but it produces here another problem; it leads to violation of unitarity in the perturbation theory.

Some authors hope that this newly produced defect is only due to the perturbative treatment of quantum gravity and that a nonperturbative procedure could lead to a renormalizable and unitary quantum theory of gravity. The methods considered in this context are sometimes criticized because they involve an essential use of expansions around flat space, so that they may be relevant only for phenomenological calculations far below Planck's energy. This critique is of course only justified if one looks for a quantum gravity theory in the sense of quantum electrodynamics; then one needs a rigorously working quantum approach. Otherwise, following our point of view, one is satisfied by approaches working consistently in some approximation, for instance, in the case of an expansion around flat space.

Let it be said once more that we confine all our considerations to genuine Einstein's GRT. Nobody can of course pretend that *gravity* must or need not be quantized, but there are arguments in favour of the conjecture that *Einstein's GRT* must not and need not be quantized rigorously. As long as we consider gravity to be described by GRT, the concept of quantum gravity (of 'gravitons') is only a formal or an approximative one, being useful to harmonize matter and gravitational equations. It should not lead to new physical (measurable) effects. As we have already mentioned and as we will latter show, this is due to the validity of the strong principle of equivalence underlying GRT.

If one had some reason to modify classical GRT in such a manner that the modified gravitational theory does not satisfy this strong principle, then one must, of course, re-discuss the total problem. Moreover, there are already a great number of gravitational theories formulated for which the concept of quantization has the same physical meaning as in electrodynamics. For instance, most bimetric theories of gravitation are of such a nature. They contain the above-mentioned supplementary 'background information' violating generally the strong principle of equivalence.

From this circumstance arises a problem which one must remember when attempting to quantize genuine (i.e., nonmodified) GRT. One should determine if such modifications (such as the R^2 terms in (1.23)) or external structures (such as a second metric) as are necessary or helpful for quantization are really pure 'quantum modifications', i.e., they do not have any consequences for classical theory. If they have such consequences then it is difficult to say that one is concerned with quantized GRT. It would be more adequate to say one has quantized an alternative gravitational theory.

Quantum theory says that one cannot distinguish between wave and particle pictures. This fact is expressed by the principle of complementarity and results in measurable quantum effects. This is true for all (nongravitational) matter and also for matter coupled to gravity; or rather, quantum theory postulates that this is true. Otherwise, using gravitational fields, one could perform experiments proposed by Einstein in his famous discussion with Bohr to falsify quantum theory or to demonstrate it as being incomplete.

However, it is quite another problem when pure vacuum gravity is concerned. Then the relation of quantum postulates to classical GRT must be discussed anew because gravity as described by GRT differs essentially from the usual matter fields by its nonlocalizability which could result in a trivialization of the indistinguishability of wave and particle pictures. The two pictures could, and this is another version of the thesis formulated above, show as different languages which are equivalent and not indistinguishable in the sense of complementarity.

To discuss the problematic relation of quantum theory and classical GRT, Rosenfeld (1966) introduced his lecture on the 1965 Einstein Symposium in Berlin with the words:

> Die Eingliederung der Gravitation in eine allgemeine Quantentheorie der Felder ist ein offenes Problem, weil zur Entscheidung der Frage nach der Quantisierung des Gravitationsfeldes die nötigen empirischen Anhaltspunkte fehlen. Hier kommt es ja nicht so sehr auf das mathematische Problem an, wie man einen Quantenformalismus für die Gravitation entwickeln soll, sondern vielmehr auf die rein empirische Frage, ob das Gravitationsfeld — und damit auch die Metrik — quantenhafte Züge aufweist. Beim Fehlen einschlägiger Beobachtungen können wir eine solche Frage nur von der erkenntnistheoretischen Seite zu beleuchten suchen, und wir dürfen dabei nicht hoffen, irgendwelche endgültigen Schlüsse zu erreichen, da erkenntnistheoretische Betrachtungen zwar dazu helfen können, die logische Struktur einer gegebenen Theorie, nicht aber deren Anpassung an die Erscheinungen zu untersuchen.*

As participants of the 1965 Einstein Symposium, we listened to Rosenfeld's lecture, and more than 20 years later we attempt here an answer to some of Rosenfeld's questions.

* The incorporation of gravitation into a general quantum theory of fields is an open problem, because the necessary empirical clues for deciding the question of the quantization of the gravitational fields are missing. It is not so much a matter here of the mathematical problem of how one should develop a quantum formalism for gravitation, but rather of the purely empirical question, whether the gravitational field — and thus also the metric — evidence quantum-like features. In the absence of relevant observations, we can only attempt to shed light on such a question from the epistemological side; and we cannot hope thereby to reach any sort of final conclusion, since epistemological considerations can indeed help in investigating the logical structure of a given theory, but not its conformity to phenomena.

Chapter 2

Quantum Mechanics and Classical Gravitation

In preparation for our study of quantum-gravity problems we shall start with a discussion of the relation between wave mechanics and General Relativity Theory. This discussion played a great role in the development of modern physics and contains, in a nutshell, a lot of the problems which arise later in quantum gravity. It refers mainly to the question of the compatibility of the quantum principle and the weak principle of equivalence, more specifically, to the compatibility of de Broglie's relation $\lambda = \hbar/Mv$ and the identity of the inertial and passive gravitational masses, $M = m$ (cf., von Borzeszkowski and Treder, 1982a, b; Treder, 1982).

The relation between general relativity and quantum theory is often discussed on the basis of *gedanken* experiments because real possibilities for experimenting in this area are lacking. Starting with a consideration of the classical Einstein—Bohr box experiment, a detailed discussion of this equation took place at the Einstein Symposium in Berlin in 1965. It was shown there that all the supposed contradictions were apparent paradoxes resulting from an inconsistent application of classical gravitational theory and quantum theory. Analyzing different *gedanken* experiments in detail, Rosenfeld (1966, 1979), for one, took this view.

The 1965 discussion was especially promoted by an example produced by Hönl, which seemed to point to a contradiction between the predictions of classical gravitational theory and those of quantum theory. Later Hönl (1981) again discussed this example to reinforce his conjecture that there is no room for quantum theory within the framework of relativity theory. According to Hönl, his example shows that quantum theory contradicts the statement of the weak principle of equivalence, according to which all bodies experience the same acceleration in an exterior gravitational field.

Following this line of argument, one is forced either to conclude that Einstein's geometrization of gravity, i.e., his own interpretation of general relativity theory, is only valid for classical physics, or, in the sense of the Einstein—Bohr discussion (Bohr, 1949; Rosenfeld, 1966, 1968) to assume that the identity of inertial and gravitational masses requires an abandonment of the quantum-mechanical complementarity of waves and particles. (Höhl himself thought that one must reformulate the theory of gravity and that wave mechanics provides corrections to the value of perihelion motion calculated in GRT.)

Since there exist experiments* showing the validity of quantum mechanics in a (homogeneous) gravitational field, some authors conclude a failure of the physical principles of general relativity theory in the domain of quantum mechanics (Hönl, 1981; Greenberger and Overhauser, 1980). However, this conclusion becomes paradoxical if it is further assumed that the general-relativistic covariant writing of the wave equations for matter fields describes the influence of gravity mathematically correctly. Indeed, Einstein's argument that this writing regards simultaneously inertial and gravitational effects gives the basis of the weak principle of equivalence. A consistent theoretical discussion should remove all apparent contradictions.

Let us say that we believe that the arguments given by Rosenfeld are generally correct; they demonstrate that, in principle, general relativity and quantum theory cannot refute one another. We emphasize that this fact results, first of all, from the principle of equivalence, which entails the innocuous nature of gravitation with regard to the field of microphysics. According to this principle, all scalar conservation laws of the special-relativistic field theory remain valid if gravitational fields are present. Indeed, since the acceleration of a physical system (as a continuous procedure) cannot change its quantum numbers (particle numbers, charges, etc.), neither therefore can a gravitational field, which is, in virtue of the equivalence principle, locally equivalent to an acceleration (Treder *et al.*, 1980).

2.1. Diffraction of Particles by a Grating

To consider the relation of the equivalence principle and quantum mechanics let us first discuss the *gedanken* experiment proposed for discussion by Bohr (1957) and Heisenberg (1930) and revived by Hönl. In this experiment, the perpendicular passage of neutral particles (neutrons) through a lattice with the lattice constant d is considered.

The velocities of all particles are assumed be equal in this experiment, while the masses, and accordingly the linear momenta, of the particles may differ from one another. According to the equivalence principle, the paths of the particles are then uniquely determined by their initial conditions, such that the deflection of the particles is independent of their masses. According to quantum theory, however, one obtains, for the deflection of particles, a diffraction pattern of the de Broglie wave which depends on the wavelength $\lambda = \hbar/Mv$ and, therefore, on the mass of the particles. If the particles have

* The first ideas for such experiments stem from Collada, Overhauser, and Werner (1972).

different masses (and equal velocities), then the lattice sorts the particles via diffraction, with respect to their masses such that a mass spectrum arises.

Indeed, let us consider two sorts of (noninteracting) particles with velocity u and the masses M_1 and $M_2 = M_1 + \delta M$, respectively, which pass through a one-dimensional lattice. Then, due to quantum theory (von Laue equation), two diffraction patterns, consisting of bands induced by particles 1 and 2, will occur on a screen. The particles of type 1 induce bands at a distance

$$H_1 = \frac{D\lambda_1}{d} = \frac{D\hbar}{p_1 d} = \frac{D\hbar}{M_1 u d} \tag{2.1}$$

from each other, and the particles of type 2 induce bands with a distance

$$H_2 = \frac{D\lambda_2}{d} = \frac{D\hbar}{(p_1 + \delta p)d} \approx H_1 - H_1 \frac{\delta p}{p_1}, \tag{2.2}$$

where D denotes the distance between lattice and screen, d is the lattice constant, and λ_1 and λ_2 are the de Broglie wavelengths of particles 1 and 2, respectively. The diffraction pattern of particle 2 is, with regard to the pattern of particle 1, shifted by a distance $|\delta H| = H_1 \delta p/p_1$. A sorting of the particles via their diffraction patterns is possible if the displacement is small enough to avoid blurring, i.e., if the inequalities $|\delta H| < H_1$ and, accordingly, $\delta p/p_1 < 1$ are satisfied.

The question raised by Hönl is as follows. Does the measurement of such a sorting effect contradict the principle of weak equivalence, according to which the deflection of particles must not depend on their masses?

Reviving an argument given earlier to discuss the Einstein—Bohr box experiment (Treder, 1971), we shall show however that the measurement of such a quantum effect does not refer in a specific manner to the relation of gravitational theory to quantum theory.

To this end, we assume that the particles and the lattice atoms are electrically charged and have such small masses that there are no gravitational, but only electromagnetic interactions between the lattice and the particles. Furthermore, the incoming particles of types 1 and 2 may have, notwithstanding their different masses, equal specific charges q_1/M_1 and q_2/M_2. If the particles are moving with a sufficiently small velocity such that their total energy is essentially given by their rest masses M_1 and M_2, then an equivalence principle holds. Both particle types are influenced in the same manner by the lattice; they both 'fall' in the potential of the electrostatic lattice with the same acceleration. Otherwise, quantum theory leads, in agreement with the above-described diffraction pattern, again to a sorting of the particles according to their masses M_1 and M_2.

Therefore, the Hönl experiment cannot refer to the specific relation between quantum theory and classical gravitational theory; accordingly, it cannot provide a contradiction between these theories. Since, in the special case discussed there, the same arguments are also correct for the relation between quantum theory and classical electrodynamics, the occurrence of a genuine contradiction would also signal a contradiction between electrodynamics and quantum theory.

This line of argument is completely analogous to the arguments given in the discussion of the Einstein—Bohr box experiment (cf. Section 2.4).

In the experiment discussed here one finds a similar situation, and one may add that a contradiction between quantum theory and gravitational theory (or electrodynamics) cannot occur, because one has to distinguish diffraction from scattering effects. The logical independence of diffraction phenomena and potential scattering becomes evident from the consideration of a light beam traversing a massive body. Due to gravitational scattering, the beam is deflected by the angle $\gamma = 4\pi\mathcal{M}G/rc^2$ (G denotes the Newtonian gravitational constant, $\mathcal{M}$ is the mass of the central deflecting body, and r is the distance from the mass centre). This deflection is, in accordance with the equivalence principle, independent of the mass (energy) of the light particles.* On the other hand, the deflecting body is also a screen for the light such that it can generate a diffraction pattern which is independent of the physical nature of the screen. (For massive particles, this phenomenon can, of course, only be described on the level of quantum theory.)

2.2. Diffraction of Particles by a Gravitational Grating

The Galilei—Newtonian particle mechanics asserts that the paths of force-free particles do not depend on their inertial masses, but only on the coordinates x^i and the velocities $\dot{x}^i = u^i$. The weak equivalence principle maintains moreover that this independence of mass is also valid if there is an exterior gravitational field acting on the particles. Otherwise, wave mechanics says that the de Broglie wave length λ of a particle possessing inertial mass M is inversely proportional to M so that, for the non-relativistic case, where $u^2 \ll c^2$, one has

$$\lambda = \frac{\hbar}{Mu} = \frac{\hbar}{p}, \quad (p^i = Mu^i). \tag{2.3}$$

The compatibility of Galilei's law and de Broglie's wave mechanics is assured by Heisenberg's uncertainty relation stating that the velocity u^i and the position x^i of a particle may only be determined with an accuracy given by the inequality relation,

$$\Delta x \cdot \Delta u \gtrsim \hbar/M. \tag{2.4}$$

* Potential scattering can, of course, be treated in both the classical and the quantum pictures. Bohr's correspondence principle just guarantees that the quantum formulas for scattering provide, for large quantum number, the classical formulas fulfilling the equivalence principle.

Accordingly, Galilei's mass-independence of the free-force motion (and of a gravitationally accelerated motion, respectively) is only established within a domain having a minimal extension determined by relation (2.3). Therefore, no contradiction arises between wave mechanics and the equivalence principle. This principle holds as well in wave mechanics as in classical-particle dynamics. Einstein thus could demonstrate the equivalence of a uniformly accelerated reference system with a resting system under the influence of a homogeneous gravitational field by means of typical wave-optics effects: aberration and Doppler shift. Einstein's *gedanken* experiments may be transferred, *mutatus mutandi*, to quantum mechanics.

To illustrate the compatibility of equivalence principle and quantum mechanics, let us turn to the above-mentioned *gedanken* experiment.

Considering this experiment to discuss the validity of the equivalence principle, one must of course presuppose a gravitational interaction between the particle and lattice, or the existence of an exterior gravitational field. Then the velocity of the particles is at least partly caused by the gravitational field. It is now essential to consider a point stressed by Rosenfeld (1966). If one simply replaces the lattice by a screen with two holes and places a mass close to one of the holes, then this mass causes by additional deflection a displacement

$$\delta H = D\delta p/p \tag{2.5}$$

(δp is the change in momentum parallel to the screen, say, in the z direction). To prevent a complete blurring of the diffraction pattern one must require that δH should be smaller than the band interval H, i.e.,

$$\delta H = D\,\frac{\delta p}{p} \leqslant H = \frac{D\hbar}{pd}\,. \tag{2.6}$$

Regarding now Heisenberg's uncertainty relation

$$\delta p \cdot \delta z \gtrsim \hbar \tag{2.7}$$

one obtains from (2.4)

$$\delta p \leqslant \frac{\hbar}{d}\,, \tag{2.8}$$

i.e., $\delta z \gtrsim d$. Accordingly, to obtain a pattern, the gravitational field has to be considered as homogeneous within the accuracy limits given by the Heisenberg relation.

We shall therefore consider the case where a system of particles moves perpendicularly to the screen or where there acts a homogeneous gravitational field in the same direction. The equivalence principle says that the motion of the system with an acceleration g has the same influence on the system as the action of the gravitational field strength $-g$. In the first case, one must, in formula (2.1), simply add the velocity v to u, where v is given, for $v^2 \ll u^2 \ll c^2$, by

$$v \approx gt = g\frac{y}{u}. \tag{2.9}$$

In the second case, one must add to u the velocity v which may be calculated from the energy conservation law

$$Mc^2 + \tfrac{1}{2}Mu^2 - M\phi = Mc^2 + \frac{M}{2}(u+v)^2$$

$$(M = \mathcal{M} = m) \tag{2.10}$$

with the gravitational potential $\phi = -gy$. From (2.10) there follows of course the same velocity

$$v \approx g\frac{y}{u} \approx -\frac{\phi}{u}. \tag{2.11}$$

The fact that v does not depend on the mass of the particles results of course from our assumption $M = \mathcal{M} = m$ and is just the statement of the weak principle of equivalence. With respect to quantum mechanics, the question arises whether there exist effects which contradict this classical independence of mass. At first sight, this is not excluded by the aforementioned fact that v does not depend on mass. Indeed, there still remains the dependence of λ on mass which is, in virtue of (2.9) and (2.11), only modified by the homogeneous gravitational field in the following manner:

$$\lambda' = \frac{\hbar}{M(u+v)} \approx \lambda\left(1 - \frac{2\phi}{u^2}\right)^{-1/2} \approx \frac{\hbar}{Mu}\left(1 - g\frac{y}{u^2}\right). \tag{2.12}$$

However, considering the pure gravitational effect, namely the Doppler shift

$$\frac{\Delta\lambda}{\lambda} \approx \frac{\phi}{u^2} \approx -\frac{gy}{u^2}, \tag{2.13}$$

it becomes obvious that the gravitational influence on the quantum system under consideration is in accordance with the weak principle of equivalence. This shift of the de Broglie wavelength corresponds to the Einstein shift of the frequency ν

$$\frac{\Delta\nu}{\nu} \approx \frac{gy}{c^2} \approx -\frac{\phi}{c^2} \tag{2.14}$$

resulting from the general-relativistic time-dilatation.

Considering now an acceleration in the z direction, i.e., parallel to the lattice, one obtains the aberration already mentioned (cf. formula (2.5)):

$$\delta H = D \frac{w}{u} = D \frac{\delta p}{p} \approx -D \frac{\delta \lambda}{\lambda} \tag{2.15}$$

(v denotes the velocity in the z direction). This displacement is even independent of mass. Calculating w in the same manner as above and inserting this into (2.14) one obtains

$$\delta H \approx 2D \frac{gz}{wu} \tag{2.16}$$

and

$$\frac{\delta \nu}{\nu} \approx \tfrac{1}{2} \frac{Mw^2}{Mc^2} \approx \frac{gz}{c^2} \approx -\frac{\phi}{c^2}. \tag{2.17}$$

Summarizing the conclusions which may be drawn from the discussion of this *gedanken* experiment, one can state that the equivalence principle holds as well in classical as in wave mechanics. The relative entities Doppler shift and Einstein shift and the aberration do not depend on mass. The quantum-mechanical mass-dependence of the absolute values of the matter-wave-lengths λ however results from the fact that, due to Heisenberg's relation, the product of the uncertainties Δu and Δx is inversely proportional to the masses of the particles. The complete mass-independence for force-free and gravitational motions is accordingly only established within a measurement interval whose lower limits are given by Heisenberg's relation.

We remark finally that a straightforward extension of the above-given arguments to relativistic velocities shows that for neutral, spinless particles with non-vanishing rest mass, the relativistic Schrödinger equation (Klein–Gordon equation) must be replaced by Fock's general-relativistic wave equation. The latter describes correctly the influence of gravitational on the matter field. Einstein's covariance principle results, for a spinless matter field, directly from the principle of equivalence; it is the mathematical formulation of this physical principle.

2.3. Gravitational Atomic Model

The fact that gravitational motion, satisfying the principle of equivalence, does not contradict quantum mechanics, within a domain of the size $\Delta p \cdot \delta x \geqslant \hbar$, also becomes obvious from the discussion of the gravitational atomic model.

The Kepler problem of classical (and approximately also of relativistic) gravitodynamics leads to ellipses, whose major axes a do not depend on the mass M or m of the planets. Indeed, in the case of a circular motion the energy E is given by

$$E = \frac{Mv^2}{2} - \frac{G\mathcal{M}m}{a}, \tag{2.18}$$

and the angular momentum is conserved,

$$ma^2\dot{\varphi} = \text{const.} \tag{2.19}$$

(M is again the inertial, m the passive gravitational mass; $v = a\dot{\varphi}$ denotes the velocity, $\dot{\varphi}$ the angular velocity, a the radius of the circular path, $\mathcal{M}$ the mass of the central body, and G the Newtonian gravitational constant.) From (2.18) then one obtains, via the virial theorem

$$-2E = Mv^2 = \frac{G\mathcal{M}m}{a} \tag{2.20}$$

and the weak principle of equivalence $M = m$, the above-stated mass-independence of a:

$$a = \frac{G\mathcal{M}}{v^2}. \tag{2.21}$$

It should be stressed that this mass-independence of a is a typical feature of gravitational theory. Differently, in electrodynamics one finds, e.g., for the motion of a (classical) electron of charge $-q$ and inertial mass M in the electrostatic Coulomb potential Q/a of a central body:

$$E = -\frac{qQ}{2a},\ a^3\dot{\varphi}^2 = \frac{qQ}{M},\ a = \frac{qQ}{Mv^2}. \tag{2.22}$$

Here q/M is no universal constant, so that one finds a q/M-dependence of the major axis a.

The situation, however, changes drastically if one considers the quantum-mechanical Kepler problem (the gravitational Bohr atomic model), i.e., the spherically symmetric solutions of the Schrödinger equation. Here one obtains with Newton's potential $U = Gm\mathcal{M}/a$ for the energy of the 'planets':

$$E_n = -\frac{1}{2}\frac{G^2m^3\mathcal{M}^2}{n^2h^2} = -\frac{1}{2}mc^2\left(\frac{Gm\mathcal{M}}{nhc}\right)^2$$
$$= -\tfrac{1}{2}mv_n^2 \tag{2.23}$$

(n denotes the natural numbers $\geqslant 1$) or

$$E_n = -\frac{Gm\mathcal{M}}{2a_n} = \frac{1}{2n^2}\frac{Gm\mathcal{M}}{a_B},\quad a_n^2\dot{\varphi} = \frac{nh}{m},$$

$$v_n = \frac{Gm\mathcal{M}}{nh}, \tag{2.24}$$

where

$$a_n = n^2 \frac{\hbar^2}{Gm^2\mathcal{M}} = n^2 a_{\mathrm{B}}, \quad \left(a_{\mathrm{B}} = \frac{\hbar^2}{Gm^2\mathcal{M}} \right) \tag{2.25}$$

defines the Bohr radii. From (2.18) and (2.25) it is now obvious that the radii a_n and the velocities v_n of the planets depend on their mass m; and it holds that

$$nv_n \cdot a_n = \hbar$$

(Bohr's quantum condition). However, one must not draw the conclusion that this fact contradicts the principle of equivalence: The Bohr principle of correspondence, guaranteeing that the quantum-mechanical version of the third Kepler law (2.22) corresponds for great quantum number n to the classical version of this law, prevents this contradiction. Bohr's principle here appears as an expression of the equivalence principle. It states that, within the limitations on the accuracy of measurement given by Heisenberg's uncertainty relation

$$\Delta(mv_n), \Delta a_n \gtrsim h, \tag{2.26}$$

for the one-body problem of the Schrödinger equation with Newton's potential, the major axes a_n of the 'planets' do not depend on the planetary masses m.

2.4. Equivalence Principle and Heisenberg's Fourth Relation*

Questioning the general validity of Heisenberg's fourth uncertainty relation $\Delta E \cdot \Delta T \gtrsim \hbar$, Einstein maintained that it is possible, on the basis of the equivalence principles, to determine the energy content E of a box through weighing by means of a spring balance; this requires a measurement of the corresponding extension q of the spring (which, for purposes of simplicity, is assumed to obey Hooke's law with a spring constant α). The box has a hole in its side, which Einstein supposed can be opened or closed by a shutter that is governed by a clock. At two definite and predetermined times T_1 and T_2 the shutter is automatically opened and then closed. If during the time $T = T_2 - T_1$ any particle escapes from the box and thereby changes its energy content, then, according to Einstein, the precision of the determination of the time of this change is evidently independent of the precision with which one weighs the energy content. Hence, this instant in time may be determined with an accuracy that is not limited by the Heisenberg uncertainty relation.

To this reasoning Bohr raised the objection that the reading of the length

* Here we follow Treder (1971).

q of the spring is possible only with an uncertainty Δq given by

$$\Delta q \gtrsim \frac{\hbar}{\Delta p}, \tag{2.27}$$

and the minimum uncertainty Δp in the momentum of the box 'must obviously ... be smaller than the total impulse which, during the whole interval T of the balancing procedure, can be given by the gravitational field to a body of mass Δm.' Hence,

$$\frac{\hbar}{\Delta q} \sim \Delta p < Tg\Delta m = Tgc^{-2}\Delta E, \tag{2.28}$$

wherein g denotes the acceleration due to gravity. Bohr concludes: "The greater the accuracy of the reading q of the pointer, the longer must, consequently, be the balancing interval T, if a given accuracy Δm of the weighing of the box with its content shall be obtained."

Now, from general relativity theory, a displacement of the box by an amount Δq in the direction of the gravitational field causes a change ΔT in the clock reading in accordance with Einstein's formula for the red shift in a gravitational field:

$$\Delta T = Tgc^{-2}\Delta q. \tag{2.29}$$

Therefore, from the uncertainty Δq of the position reading, on the one hand, and the condition (2.28), on the other, precisely the Heisenberg uncertainty relation

$$\Delta E \Delta T \gtrsim \hbar \tag{2.30}$$

is obtained.

The critical element in Bohr's argument is, in our opinion, the relationship between the uncertainty in the momentum of the box and the duration T of weighing. We shall return to this point later. At present we want to show, through a somewhat modified Einstein experiment, that Bohr's assumption, namely that the consideration of the time dilatation (2.29) provides the explanation of the Einstein box experiment, cannot be correct. We assume that the box is, for practical purposes, massless and contains indistinguishable electrically charged particles, say, protons*, which may be at rest with respect to one another and to the observer. Furthermore, the protons may be sufficiently far apart so that the total energy E of the box is given by the sum of the equilibrium energies of the individual protons: $Mc^2 = \Sigma\, m_p c^2 = E$. Since all protons possess the same specific charge e/m_p, an equivalence principle holds: in the situation described here, it expresses the

* One can also replace the protons by suitable macroscopic particles, e.g., by equal-sized charged particles or bubbles, where — as in the Millikan experiment — each drop contains exactly one elementary charge. In this case, the determination of the mass through 'electrical weighing' is feasible not only in the sense of a *gedanken* experiment.

proportionalities between the total charge Q, the total energy E, and the total mass M:

$$Q = \frac{e}{m_P} M = \frac{e}{m_P} \frac{E}{c^2}. \tag{2.31}$$

We suspend the box, prepared in this manner, in a homogeneous electrostatic field which is produced, for example, by a negatively charged plane surface. The field strength of this field is then given by $\mathbf{F} = (F, 0, 0)$, and we find a proportionality of the spring's elongation q to the total charge Q of the box:

$$aq = QF \tag{2.32}$$

On account of (2.31), this produces also a proportionality between the spring elongation and the energy content of the box,

$$q = \frac{eF}{am_P c^2} E. \tag{2.33}$$

If corresponding application of Bohr's considerations are made here, we obtain the relationship between the uncertainties Δp, Δq, and ΔE (of the box-momentum, -elongation, and -energy), on the one hand, and the duration T of the weighing process, on the other:

$$\frac{\hbar}{\Delta q} \sim \Delta p < TF\Delta Q = \frac{e}{m_P c^2} TF\Delta E. \tag{2.34}$$

Here there is, however, no longer any Einstein time dilatation, that is, the uncertainty Δq of the elongation causes no uncertainty ΔT of the time interval T.

Hence, we are able to imagine that, during the time the box is opened, some protons escape from the box (with very small velocities), where for weighing in a homogeneous electrostatic field, no relationship exists between the uncertainties of the time and energy measurements. Thus, if Bohr's vindication of Heisenberg's fourth uncertainty relation proves to be correct for a gravitational field, then the uncertainty relation must be invalid for measurements in an electrostatic field.

In fact, however, Bohr's estimate of the uncertainty of the spring's elongation is not understandable, for the uncertainty of the box-momentum is brought about by the intervention of the observer in his attempt to read the spring length q, and this disturbance by the observer has nothing to do with the effect of the gravitational field (or electrostatic field) upon the box. Hence, no connection can exist between the duration of the weighing and the disturbance of the momentum. It is also, in fact, not understandable why the momentum should become more imprecise with increasing weighing time. The Einstein question consequently must be pondered anew, since Bohr's attempt to push it *ad absurdum* has not met with success.

Einstein's assumption, namely, that it is possible to determine at precise points in time the initial as well as the final energy, is of course correct. The question is only how close to one another the measurement times can be moved. Certainly, as has become evident through the discussions of Bohr and Heisenberg, the relation $\Delta E \Delta T \gtrsim \hbar$ does not at all assert a connection between the uncertainties of energy and time measurements upon stationary systems; the relation indicates rather that a time interval $\Delta T \gtrsim \hbar/\Delta E$ must separate the first and second measurements in a determination of an energy difference $E_1 - E_2$ with an uncertainty ΔE. Consequently, the fourth uncertainty relation has direct significance only for nonstationary systems, in that it connects the lifetime (or half-life) ΔT of the nonstationary state with its spread in energy (linewidth), such that if the initial condition of the nonstationary system is precisely given, then the energy of the final state exhibits a linewidth

$$\Delta E \sim \hbar/\Delta T. \tag{2.35}$$

Let us consider an ensemble Σ of certain quasi-stationary systems which initially possess a sharply defined energy E_1. Then, in the final state, the energies of the systems constituting the ensemble will be statistically distributed with a half-width given by (2.35). The final energy E_2 of each individual system is precisely measurable; but the systems possess different final energies, despite their having the same initial energy. (They have then also given off different energy quanta.)

Such a quasi-stationary state now also leads to the uncertainty relation for the Bohr—Einstein box. Let us consider an ensemble of these boxes, all possessing at the start the same spring tension and total energy. Before the emission of a particle from the box occurs, the system 'box + spring' is in a state of stationary equilibrium. At the instant that a particle of mass $\mathrm{d}m$ is emitted, the system 'box + spring' does not at first achieve a stable condition, because the spring tension and the gravitational force no longer balance one another. The system begins to vibrate, and indeed with an amplitude $\mathrm{d}q$ proportional to the change of the force acting upon the spring

$$\alpha\, \mathrm{d}q = g\, \mathrm{d}m; \tag{2.36}$$

and it becomes in this manner a quasi-harmonic oscillator with the vibrational energy

$$\tfrac{1}{2}\, \alpha (\mathrm{d}q)^2 = g\, \mathrm{d}m\, \mathrm{d}q. \tag{2.37}$$

As long as the box vibrates up and down, the change $\mathrm{d}q$ of the spring elongation and, therefore, $\mathrm{d}m = c^{-2}\, \mathrm{d}E$ are not readable. The system is, however, not a true harmonic oscillator; rather, the vibrations are damped, while in the spring, through the action of internal friction, and the vibrational energy (2.37) is converted into heat energy, which finally leaves the system through heat conduction or radiation. This heat transfer is a dissipative process in which information is unavoidably lost. The energy carried off is not precisely determined; rather, it spreads statistically, so that each system of the ensemble suffers a slightly different energy loss. Leaving details aside, it follows from the general characteristics of damped vibrations that the fluctuation ΔE of the total amount of dissipated energy of the ensemble members are inversely proportional to the median lifetime ΔT of the oscillation state: $\Delta E \sim \hbar/\Delta T$.

After complete cessation of the vibrations, it is possible to read off accurately the spring elongation and hence the energy content of the system. The individual systems of the ensemble have, however, in the final state somewhat different energies, as each of them has lost a different amount of energy in the damping of its vibrations. Consequently, the different systems of this ensemble will not settle down in the steady state with equal elongations; these will spread. Corresponding to the different energy contents of the boxes, we have

$$a\Delta q = gc^{-2}\Delta E = \frac{g\hbar}{c^2\Delta T} . \tag{2.38}$$

The spread Δq is larger, the shorter the average lifetime of the quasi-stationary vibrational state, and the connection between lifetime and energy spread is given by the Heisenberg uncertainty relation.

Einstein's paradox may thus be resolved in the following manner. With the emission of a particle from the box, the Einstein system 'box + spring' goes over into a quasi-vibrational state and through damping of these oscillations evolves into the final, steady state. The energy of the final state scatters from one experiment to another, and the spread ΔE is larger, the shorter the time that the system requires on the average for attaining the final condition. Therefore, in the process of measuring the energy content, a time interval of at least $\Delta T = \hbar/\Delta E$ must separate the initial and final states if the determination of the final state is to result with a reproducible accuracy ΔE. The fulfilment of the fourth uncertainty relation is therefore a consequence of the quantum properties of the spring, and has nothing to do with any connection between quantum theory and gravitation theory.

To support Bohr's opinion that the weak principle of equivalence must be assumed to guarantee the validity of the fourth Heisenberg uncertainty relation, Costa de Beauregard (1985) modified the Einstein *gedanken* experiment for the limit of free fall.

However, according to the weak principle of equivalence, a freely falling body has no weight at all! The question raised by Costa de Beauregard is answered by the same arguments which were used with respect to the spatial components of Heisenberg's uncertainty relation in Section 2.2.

With Galileo's fall acceleration g, one has for the momentum p of a freely falling body of mass m (in z direction)

$$p = mv = m(gt + v_0),\ v_0 = \text{const}, \tag{2.39}$$

where t is the fall time. The fall path amounts to

$$z - z_0 = \tfrac{1}{2} gt^2 + v_0 t,\ z_0 = \text{const}. \tag{2.40}$$

In the same classical approximation, the proper energy of the freely falling mass is given by

$$E = E_0 + \frac{m}{2}(gt)^2 + mgtv_0,\ E_0 = \text{const}. \tag{2.41}$$

Accordingly, one obtains for the uncertainties Δp, Δz, ΔE, and Δt of p, z, E, and t:

$$\Delta p = m\Delta v = \Delta m \cdot g \cdot t, \tag{2.42}$$

$$\Delta z = gt\,\Delta t + v_0 \Delta t, \tag{2.43}$$

$$\Delta E = mv\Delta v = mg^2 t\,\Delta t + mgv_0 \Delta t. \tag{2.44}$$

The z component of Heisenberg's uncertainty relation reads

$$\Delta p \Delta z = m\Delta v \Delta z = mg^2 t(\Delta t)^2 + mgv(\Delta t)^2 \geqslant \hbar \tag{2.45}$$

and this leads really, together with (2.44), to the fourth uncertainty relation

$$\begin{aligned}\Delta E\,\Delta t &= mg^2 t(\Delta t)^2 + mgv_0(\Delta t)^2 \\ &= \Delta p \Delta z \gtrsim \hbar.\end{aligned} \tag{2.46}$$

2.5. Quantum Mechanics and the Weak Principle of Equivalence

Summarizing our short discussion of the behaviour of quantum-mechanical systems in exterior non-quantized gravitational fields, one may state that, for two reasons, there arise no contradictions between quantum mechanics and classical gravitational theory. First, by virtue of the weak principle of equivalence, gravity is physically harmless in microscopic regions. Second, Bohr's principle of correspondence guarantees that the transition from solutions of the Schrödinger equation (with the Newtonian potential) to macroscopic regions must be performed so that the fundamental principle of gravitational theory (the equivalence principle) is satisfied.

The first point is especially interesting because it does not only say that contradictions do not arise, but it even implies that both theories support one another. Indeed, the fact that homogeneous exterior gravitational fields do not change the quantum numbers of a quantum system such as a caesium atom renders it just possible that atomic clocks being logically independent of the gravitational theory, can be used for time measurement in gravitational theory. (This is a condition necessary for a physical interpretation of each theory, in particular also of the gravitational theory.) On the other hand, gravitational theory with 'gravitation-free' or, in GRT, 'uncurved' microscopic regions allows us to formulate the usual quantum mechanics. Only quantum mechanics together with GRT can form 'complete physics'.

The relation between quantum mechanics and Einstein—Newtonian gravitational theory is so constituted that the validity of the weak principle of equivalence is assured. This implies however that quantum mechanics and gravitational theory are not genuinely unified in one theory. They are two separate parts of the fundaments of physics, whose compatibility is assured by Heisenberg's uncertainty relation. Accordingly, one can here already formulate the conjecture that the consideration of full GRT, which satisfies a stronger version of the principle of equivalence (Einstein's principle) than the one governing the motion of masses in an exterior gravitational field, requires the validity of stronger uncertainty relations than Heisenberg's. Otherwise, General Relativity and quantum theory should not be compatible.

Chapter 3

Measurement in Quantum Gravity

In discussing problems of quantum mechanics and quantum field theory, it has been repeatedly mentioned that, following von Neumann's (1955) axiomatization and that of other authors, the measuring process is reduced to rules which bring about only a formal relationship between certain elements of quantum formalism and certain states of the measuring instrument. It is sometimes stressed, first, that one should follow Bohr and Heisenberg and analyse the interaction between the microscopic object system and the measuring instrument and, second, that the macroscopic nature of the measuring device should be taken into consideration (cf., e.g., de Muynck, 1984).

To our minds, one has however to go a step further. We think that one has to follow precisely the Bohr—Rosenfeld paper (Bohr and Rosenfeld, 1933). The line of arguments given there does not only show that one has to consider the interaction between a microscopic object system and macroscopic measurement device, but it also specifies the term 'macroscopic measurement device'. It seems to us to be a lack of many investigations that the Bohr—Rosenfeld definition of this concept is not taken seriously enough.

It is especially necessary to go back to the Bohr—Rosenfeld analysis if one is interested in the foundation of quantum gravity. Discussing quantum gravity means dealing with a subject for which neither the physical meaning of quantization nor the mathematical tools for quantization are satisfactorily clear.* In such a field it would be dangerous to work on the basis of a formal relationship founded for other physical cases, e.g., for electromagnetism. Such an approach would be reasonable if the fundamental physical problems were solved so that one had merely to elaborate the details of the theory. If one wants, however, to found quantum gravity or to find out the physical status of the approaches to quantum gravity proposed to date, then one has to discuss the measuring process. It is the only guideline to a unification of quantum theory and GRT. Indeed, as long as one does not arrive at a satisfactory quantum theory of a field, the measurement discussion regarding the interaction between the object system and the measuring instrument just

* For a discussion of the present state of quantum gravity, cf., e.g., Asthekar (1983).

provides the rules for transferring quantum mechanical laws of the 'quantum-mechanically handled' apparatus to the field. On the other hand, if one has some quantum field formalism, one can investigate its physical significance by such a measurement discussion. In both cases the measurement discussion is relevant, and it is quite clear that this discussion is strongly dependent on the presupposed structure of the measurement body. It is not sufficient to say it is macroscopic. For a real interaction discussion it must be specified in more detail.

Therefore, one has to do this measurement discussion for gravitational theory. More precisely, for GRT one has to repeat the measurement discussion which was done by Bohr and Rosenfeld for the electromagnetic field. One cannot simply take over the electromagnetic results, because according to Einstein (see Heisenberg, 1969), the physical theory describing a system itself tells which entities are measurable.

The guideline of the necessary measurement discussion was formulated by Bohr (1957) as follows:

Trotz aller Unterschiede in den physikalischen Problemen, die zur Entwicklung der Relativitätstheorie und der Quantentheorie Anlaß gegeben haben, enthält ein Vergleich der rein logischen Aspekte relativistischer und komplementärer Darstellungsweise weitgehende Ähnlichkeiten hinsichtlich des Verzichtes auf die absolute Bedeutung althergebrachter physikalischer Attribute der Objekte. Auch die Vernachlässigung der atomaren Konstitution der Meßgeräte selber bei der Beschreibung tatsächlicher Erfahrungen ist gleich charakteristisch für die Relativitäts- und Quantentheorie. Die Kleinheit des Wirkungsquantums verglichen mit den Wirkungen, um die es sich bei gewöhnlichen Erscheinungen einschließlich Aufstellung und Bedienung physikalischer Apparate handelt, ist in der Atomphysik genau so wesentlich wie die riesige Anzahl von Atomen, aus denen die Welt besteht, in der Allgemeinen Relativitätstheorie, welche bekanntlich verlangt, daß die Dimensionen der zur Winkelmessung benutzten Geräte klein gegen den Krümmungsradius des Universums gemacht werden können.*

3.1. The Bohr–Rosenfeld Principles of Measurement in Quantum Field Theory

(a) *The Landau–Peierls Arguments*

After a long period of resistance, the photon hypothesis was accepted by the

* "Despite the difference in physical problems, which gave rise to the development of relativity theory and quantum theory, a comparison of the purely logical aspects of relativistic and complementary kinds of representation shows far-going similarities with respect to a renunciation of the absolute meaning of old physical attributes of objects. Also the neglect of the atomistic constitution of the measurement apparata themselves in describing real experiences is as characteristic of relativity as of quantum theory. The smallness of the quantum of action compared with the actions concerning usual phenomena, including the establishment and servicing of physical apparata, is in atomic physics as essential as the huge number of atoms of which our world consists, is in general relativity theory, requiring — as is well known — that the dimension of the measurement equipment used for angle measurement can be made small compared to the curvature radius of universe."

majority of the physics community in the late 1920s.* At that time, even Max Planck was ready to defend this thesis, and he asked for a self-consistent quantum-electromagnetic theory (cf. Planck, 1927).

This theory then was established by Heisenberg, Jordan, and Pauli (Jordan and Pauli, 1928; Heisenberg and Pauli, 1929, 1930). Otherwise, one had yet to solve a variety of problems, and it was not quite clear in the early thirties, to what extent one should consider quantum electrodynamics as a new physical theory. The standpoint that it is only a (yet incomplete) mathematical formalism without any new physical consequences could not be excluded from consideration at that time. That point was especially stressed by Landau and Peierls (1931) when they published a paper, wherein they attempted to demonstrate that new uncertainty relations arise which impose limitations on the measurability of electromagnetical field quantities.

Landau and Peierls argued as follows. These new limitations imply that no predictable measurements exist for quantities which characterize the quantized electromagnetic theory. The applicability of this theory should, therefore, be restricted to processes where the state of the system varies sufficiently slowly. In cases in where the ordinary Schrödinger equation is applicable, i.e., in non-relativistic approximation cases, the validity of quantum theory is of course always true. For radiation alone, quantum theory is however never meaningful, since the limit $c = \infty$ then has no sense.

In other words, Landau and Peierls pretended to show that, contrary to Einstein's photon hypothesis, electromagnetic fields themselves need not be quantized because quantization does not result in measurable effects. Moreover, quantization generates a lot of technical difficulties, e.g., divergencies, so that it is an unhappy procedure.

Today one knows of course that these arguments must be wrong somewhere, because there exists a physically meaningful quantum electrodynamics. Nevertheless, for a better understanding of the Bohr—Rosenfeld reply to Landau and Peierls, being essential for clearing the status of quantum gravity, we shall outline here some of the Landau—Peierls arguments. This is especially useful because we shall demonstrate below (cf. Sections 3.1b, 3.2) that, in accepting Bohr's and Rosenfeld's objections to the Landau—Peierls paper, the conclusion which Landau and Peierls wanted to draw for electromagnetism can really be drawn for gravity.

Landau and Peierls started with a consideration of momentum measurement in the relativistic case. Taking into account that the velocity v of a particle cannot exceed vacuum light velocity c, Heisenberg's uncertainty relation

$$\Delta p_x \Delta x \gtrsim \hbar \tag{3.1a}$$

* For a detailed representation of the history of the photon hypothesis, cf. A. Pais (1979).

provides

$$\Delta p_x \Delta t \gtrsim \hbar/c \tag{3.1b}$$

(p_x is the x component of the momentum $\mathbf{p}$ which is measured, Δt denotes the time necessary for the momentum measurement).

It was assumed then that, in measurements with a charged body, in addition to the uncertainties (3.1), a further perturbation of the measurement arises because the body will emit radiation during the necessary change of velocities.

Presupposing that one realizes an arrangement where the velocity of the body before and after the measurement is considerably small compared with c, the non-relativistic formula for radiation damping can be used. The energy emitted is then

$$\Delta E = \frac{Q^2}{c^3} \int \dot{v}^2 \, \mathrm{d}t \sim \frac{Q^2}{c^3} \frac{(v'' - v')^2}{\Delta t}, \tag{3.2}$$

where Q is the charge of the measurement body and v' and v'' are the velocities before and after the measurement. This unknown change of energy leads to the additional uncertainty

$$\Delta p_x (v'' - v') > \frac{Q^2}{c^3} \frac{(v'' - v')^2}{\Delta t} \tag{3.3a}$$

or

$$\Delta p_x \Delta t > \frac{Q^2}{c^3} (v'' - v'). \tag{3.3b}$$

Landau and Peierls stressed that for electrons, where $Q = e$, this inequality gives no new information since even in the most unfavourable case, where $v' = v'' + c$, it gives only $\Delta p_x \Delta t > e^2/c^2$ and this relation is weaker than (3.1b) since $e^2 < \hbar c$. For macroscopic bodies, however, the relation (3.3b) is relevant. Using (3.1) in the form

$$(v'' - v') \Delta p_x > \frac{\hbar}{\Delta t} \tag{3.4}$$

relation (3.3b) can also be written as follows

$$\Delta p_x \Delta t > \frac{\hbar}{c} \sqrt{\frac{Q^2}{\hbar c}}. \tag{3.5}$$

Considering now the simple method of measuring the x component F of an electric field by observing the acceleration of a charged measurement body of large mass and small velocity, Landau and Peierls used the formula

$$Q \, \Delta F \, \Delta t > \Delta p_x \tag{3.6}$$

for the ΔF accuracy of the electric field strength. Using (3.5), one obtains from (3.6) finally the uncertainty relation

$$\Delta F(c\Delta t)^2 \gtrsim \sqrt{\hbar c}\,. \tag{3.7}$$

This is the decisive inequality relation derived by Landau and Peierls. From this relation they deduced that one cannot perform measurements which show the physical existence of light quanta, because in a radiation field no measurement can be carried out with certainty within such a short time, i.e., no measurements for which every possible result gives information about the state of the system.

(b) *The Bohr—Rosenfeld Arguments*

Bohr and Rosenfeld (1933) reanalysed electromagnetic field measurements to show that, in contrast to the Landau—Peierls result, there arise the same limitations on the accuracy of fields (i.e., the same uncertainty relations for ΔF) as postulated in the formalism of quantum electrodynamics. Accordingly, calculable quantum effects can be measured, i.e., quantum electrodynamics has a physical meaning.

It is important to emphasize that Bohr and Rosenfeld did not oppose the conclusion that in the case of the validity of inequality (3.7), quantum field effects of electromagnetism were not measurable; they rather showed that (3.7) is incorrect and that therefore this conclusion cannot be drawn. Furthermore, they stressed that, if the measurability of field quantities is subjected to further restrictions which go essentially beyond the presuppositions of quantum field theory, these supplementary stronger restrictions would deprive this theory of any physical sense. The fundamental postulate which they formulated in their paper was to require that the uncertainty relations following from the measurement discussion have to be the same as the relations established by the theory (e.g., by the commutation rules of quantum field theory). Any contradiction would be a serious dilemma.

By closer consideration of the measurement process, Bohr and Rosenfeld aimed to show that in quantum electrodynamics no contradiction arises between quantum-field formalism and measurement theory as pretended by Landau and Peierls.

The main points of Bohr and Rosenfeld are the following: Measuring physical effects, one has strictly to distinguish between the system whose quantities are to be measured, the measurement device (a test body), and their mutual interaction. As far as its *constitution* is concerned, one has to consider the measurement body as a classical body. This means that one has to neglect its atomic structure and presuppose a rigid body in the sense of classical mechanics, carrying a homogeneously distributed charge. In particular, one cannot perform field measurements by an electron showing radiative back reaction, as was done by Landau and Peierls.

For a measurement arrangement one has to realize classical bodies with an arbitrarily high accuracy. The main argument for this standpoint is given by the statement that a consideration of the atomic structure of the measurement body is not adequate to a measurement of pure fields because it transfers the discussion to the level of the amalgamation of quantum-electrodynamics and atomic matter theory. If one wants to measure field effects, one has to evade such a mixture. Of course, enormous difficulties can arise in constructing such an artificial test body, but one has to try. Otherwise, the distinction between the measurement apparatus and the physical system whose parameters are to be measured would be destroyed and the concept of measurement would lose its sense. Considering, however, the *interaction* between the test body and the electromagnetic field system, one has to use the quantum-mechanical laws, in particular, the Heisenberg uncertainty relations. They impose an absolute restriction on the displacement of test bodies which cannot be 'compensated' by a refined mechanism.

Following Bohr and Rosenfeld, one has to make the following assumptions:

(i) In order to have a definite case in mind, one considers the measurement of the field average over a spacetime domain of volume V and duration T. For this purpose, we use a measurement body, whose electric charge Q is uniformly distributed over the volume V with a density ρ, and determine the values p'_x and p''_x of this body's momentum components in the x direction at the beginning t' and the end t'' of the interval T.

(ii) The time interval Δt required for the momentum measurement can be regarded as negligibly small compared to T,

$$\Delta t \ll t'' - t' = T. \tag{3.8}$$

(iii) One can neglect the displacements suffered by the measurement body due to the momentum measurement in comparison with the linear extension L_0 of the domain V,

$$\Delta x \ll L_0. \tag{3.9}$$

(iv) By choosing a sufficiently heavy test body, we minimize the acceleration given to it during the time interval T by the field so that we can disregard any radiation back-reaction.

(v) To guarantee that the measurement body behaves as a classical one during the measurement of the momentum change, we demand

$$cT < L_0. \tag{3.10}$$

The borders of the measurement body then are separated by space-like distances so that no causal relation exists between them. To exclude moreover that the measurement body does not behave classically only from the viewpoint of relativity theory but also from the viewpoint of the atomistic theory of matter, one must demand

$$uT < L_0 \tag{3.11}$$

where $u = c^2/v$ is the phase velocity. Indeed, considering the measurement body as atomistically structured, one has to assume de Broglie's relations

$$E = mc^2 = \hbar\nu \tag{3.12}$$

and

$$p = mv = \hbar/\lambda. \tag{3.13}$$

Then there result quantum correlations between different regions of this body caused by processes propagating with the phase velocity

$$u = \frac{c^2}{v} = \frac{E}{p} = \lambda\nu. \tag{3.14}$$

They are excluded by requiring relation (3.11). By virtue of $u \gtrsim c$, (3.11) is a stronger requirement than (3.10).

Assumptions (i)—(v) concern the structure of the measurement body or, more generally, the measurement arrangement. In principle, we can choose mass, charge, linear extension of the measurement body and duration of measurement so that these requirements are satisfied. The Bohr—Rosenfeld requirements imply especially the conditions

$$Q^2 \gtrsim \hbar c \tag{3.15}$$

and

$$L_0 > \frac{Q^2}{Mc^2} \tag{3.16}$$

which one has to impose on the dimension of the measurement body.

Let us consider now Lorentz' equations of motion underlying the momentum measurement here under consideration:

$$M\ddot{x}^\mu = QF_{\mu\nu}\dot{x}^\nu + QF'_{\mu\nu}\dot{x}^\nu. \tag{3.17}$$

Here $F_{\mu\nu}$ denotes the exterior field we want to measure and $F'_{\mu\nu}$ the self-field of the measurement body. Assuming point charges, the second term on the right-hand side of (3.17) was studied by various procedures (Dirac, 1938; Barut, 1974, 1979). The result is as follows. The self-field term in (3.17) can be written as a sum of two terms

$$-\frac{1}{2c^2}\lim_{\varepsilon \to 0}\frac{Q^2}{\varepsilon}\ddot{Z}_\mu + \tfrac{2}{3}Q^2\left(\frac{\dddot{Z}_\mu}{c^3} + \frac{\dot{Z}_\mu \ddot{Z}^2}{c^5}\right). \tag{3.18}$$

In our case of an extended body of finite extension L_0 moving under the influence of an external electric field F in the x direction, we obtain from

(3.18) the following order-of-magnitude relation:

$$M\ddot{x} \sim QF + \frac{Q^2}{c^2 L_0}\ddot{x} + \frac{Q^2}{c^3}\dddot{x} + \frac{Q^2}{c^5}\dot{x}\ddot{x}^2. \tag{3.19}$$

From condition (3.16) we see now that

$$M\ddot{x} > \frac{Q^2}{c^2 L_0}\ddot{x} \tag{3.20}$$

such that this term and, furthermore, the higher-order radiation corrections can be neglected.

W. Pauli was the first who criticized Landau and Peierls' assumption (3.3) taking radiative back-reaction into account. In a letter to Heisenberg, Pauli (1933) wrote that, in discussions with Peierls, he had formulated the following objection to this procedure. A charged test body can be brought into a very large opaque box. The repulsion of the box as well as the energy absorbed by the walls of the box can be measured arbitrarily exactly. Therefore, the back-reaction of radiation should not cause any unknown uncertainty.

Regarding all the assumptions made above, the average field strength F can be determined via relation

$$\delta p \equiv p_x'' - p_x' = \rho F V T \tag{3.21}$$

and the field uncertainty is

$$\Delta F = \frac{\Delta(\delta p)}{\rho V T}. \tag{3.22}$$

In the momentum measurement we also encounter conditions that are independent of the structure of the measurement body. Any measurement of the momentum component p_x is with an uncertainty Δp_x and satisfies Heisenberg's uncertainty principle (3.1a). As was emphasized by Bohr and Rosenfeld (1933), this relation concerns the handling and not the structure of the measurement body which we have to note in our list of requirements on quantum measurements:

(vi) Any measurement of the momentum component p_x is accompanied by a loss Δx of one's knowledge of the position of the body in question, whose order of magnitude is given by

$$\Delta p_x \Delta x \sim \hbar. \tag{3.23}$$

Combining the two relations (3.22) and (3.23), one obtains

$$\Delta F \sim \frac{\hbar}{\rho\, \Delta x\, V T} \tag{3.24}$$

and, due to (3.10), one finally has

$$\Delta F L_0^2 \gtrsim \hbar c/Q \tag{3.25}$$

or

$$\Delta F L_0^3 \gtrsim \frac{\hbar}{c}\frac{Q}{M}. \tag{3.26}$$

These relations differ essentially from the Landau—Peierls relation (3.7) by the fact that they do not imply a restriction on the accuracy to be achieved by field measurements because we still have the values of the charge density ρ or charge Q at our disposal. The right-hand side can be made arbitrarily small by choosing a sufficiently large value of ρ or Q.

As was mentioned above, the main argument against relation (3.7) is that it is derived by considering radiative back-reaction. Here we see that relation (3.7) also follows from (3.25), if one assumes a measurement body carrying electron charge $Q = e$. Indeed, setting $Q = e$ and using the relation

$$1 > \left(\frac{e^2}{\hbar c}\right)^{1/2}$$

one finds from (3.25) again

$$\Delta F L_0^2 \gtrsim \frac{\hbar c}{Q} > \frac{\hbar c}{Q}\left(\frac{e^2}{\hbar c}\right)^{1/2} = (\hbar c)^{1/2}.$$

Therefore, Bohr and Rosenfeld (1933) and later Rosenfeld (1966) stressed the need to use measurement bodies with $Q \gtrsim (\hbar c)^{1/2}$. Thus, only relation (3.16) together with (3.15) prevents the Landau—Peierls formula.

Before continuing our discussion, let us make a remark on modern measurement analysis (cf. von Borzeszkowski, 1985b). It starts from Feynman's path integral

$$A(x', x) = \int_{I(x, x')} \mathrm{d}\{x\} \exp\left(\frac{i}{\hbar} S\{x\}\right) \tag{3.27}$$

giving the total transition amplitude for the transition from the point x to the point x'. ($S\{x\}$ is the action integral calculated along a given path from x to x' and the integral over $\{x\}$ denotes the summation of the amplitudes for the different paths between x and x'.) If one performs now a (continuous) measurement providing some information about the path of transition (measurement value α), then one does not have to integrate over all paths $I(x, x')$ but only over a set $I_\alpha(x, x')$. Assuming that this preference of certain paths can be expressed by a weighing function

$$\rho_{\{a\}}\{x\} = \exp\left(- \frac{\langle (x-a)^2 \rangle}{\Delta a^2} \right) \tag{3.28}$$

one may write

$$A_{\{a\}}(x, x') = \int \mathrm{d}\{x\}\, \rho_{\{a\}}\{x\} \exp\left(\frac{i}{\hbar} S\{x\} \right). \tag{3.29}$$

Here $\{a\}$ denotes a path determined by the measurement and Δa the accuracy of the measurement. Analysing $A_{\{a\}}(x, x')$ for an oscillator under the influence of an external force, the uncertainty ΔK of this force was estimated. (For a detailed discussion, see Mensky (1983).) For an optimal measurement, where

$$\Delta a \approx \left(\frac{2\hbar}{m\tau\, |\Omega^2 - \omega^2|} \right)^{1/2} \tag{3.30}$$

it amounts to

$$\Delta K \gtrsim \left(\frac{m\hbar\, |\omega^2 - \Omega^2|}{\tau} \right)^{1/2}. \tag{3.31}$$

One sees that this relation is equivalent to the Bohr—Rosenfeld relation (3.25). Indeed, multiplying ΔK by Δa and considering that

$$\Delta a \approx L_0,$$

$$\tau = T \lesssim L_0/c \tag{3.32}$$

and

$$\Delta K \approx Q\Delta E \tag{3.33}$$

one obtains

$$\Delta F \Delta a \approx \Delta F L_0 \gtrsim \frac{\hbar}{\tau Q} \gtrsim \frac{\hbar c}{Q L_0}. \tag{3.34}$$

The Bohr—Rosenfeld derivation of this relation has the advantage of making it obvious that (3.31) results mainly from Heisenberg's uncertainty relation.

Before using the Bohr—Rosenfeld principles for a discussion of the physical status of quantum gravity, we shall make some comments and also summarize some points discussed by Bohr and Rosenfeld (1933). They show in particular that the inequality relations derived in the previous section are in agreement with the uncertainty relations obtained from the quantum formalism of electromagnetic field theory.

To this end, we consider the average value

$$F \equiv \overline{E}_x = \frac{1}{VT} \int \mathrm{d}t \int E_x \,\mathrm{d}v \tag{3.35}$$

of the x component of the electric field calculated over two given spacetime regions I and II, for which in quantum electrodynamics commutation rules exist and lead to uncertainty relations of the type

$$\Delta \overline{E}_x^{(\mathrm{I})} \Delta \overline{E}_x^{(\mathrm{II})} \sim \hbar (\overline{A}_{xx}^{(\mathrm{I,\,II})} - \overline{A}_{xx}^{(\mathrm{II,\,I})}), \tag{3.36}$$

where $\overline{A}_{xx}$ denotes a Green function. Using (3.26), one can now form an essential quantity for the physical interpretation of quantum electrodynamics: the square root $\mathcal{H}$ of the products, given by (3.26), of the complementary uncertainties of two field averages over spacetime regions that only partially coincide, being displaced relative to each other by spatial and temporal distances of order of magnitude L_0 and T, respectively. For field strengths that are essentially larger than $\mathcal{H}$, we enter the domain of validity of classical electromagnetical theory, where all quantum mechanical features of the formalism lose their significance. An estimate of $\mathcal{H}$ shows that for $L_0 > cT$ one has

$$\mathcal{H} \sim \left(\frac{\hbar}{L_0^3 T} \right)^{1/2} \tag{3.37}$$

and, on the other hand, in the case $L_0 \lesssim cT$,

$$\mathcal{H} \sim \frac{(\hbar c)^{1/2}}{L_0 c T}. \tag{3.38}$$

Otherwise, one can derive another critical field strength $\mathfrak{S}$, in the sense that only when considering field averages essentially larger than $\mathfrak{S}$ are we allowed to neglect the corresponding fluctuations. An estimate for $L_0 > cT$ is

$$\mathfrak{S} \sim \frac{(\hbar c)^{1/2}}{L_0^2} \tag{3.39}$$

and for $L_0 \lesssim cT$,

$$\mathfrak{S} \sim \frac{(\hbar c)^{1/2}}{L_0 c T}. \tag{3.40}$$

To investigate the quantum features of the electromagnetic theory, one has to assume $L_0 \gg cT$. Then the classical field strength $\mathcal{H}$ is much larger than $\mathfrak{S}$ and, therefore, in testing the characteristic consequences of the formalism, we can, to a large extent, disregard fluctuations. This investigation shows that

$$\mathscr{H} \sim \mathfrak{S} \sim \frac{(\hbar c)^{1/2}}{L_0 cT} \sim \frac{(\hbar c)^{1/2}}{L_0^2} \tag{3.41}$$

is a limitation arising only if we assume $L_0 \sim cT$, as was done by Landau and Peierls.

This result shows that the two universal constants vacuum light velocity c and quantum action $\hbar$ entering vacuum quantum electrodynamics are not sufficient to determine any limitations. This situation changes for each theory containing a further universal constant.

3.2. Measurements in Quantum Gravity

The discussion on the measurability of quantum effects of gravity was started some decades ago in the papers of Rosenfeld (1957, 1966), Wheeler (1957, 1964), Regge (1958), Peres and Rosen (1960), Treder (1961, 1963), and DeWitt (1962, 1964). As was shown, one can transfer the Bohr—Rosenfeld considerations to gravitational fields if one replaces the field strength F, the passive charge q and the active charge Q by the corresponding gravitational quantities. This can be done by the rules

$$F \to \frac{c^2}{G}\Gamma, \tag{3.42a}$$

$$Q \to \sqrt{G}\mathscr{M}, \tag{3.42b}$$

$$q \to \sqrt{G}m, \tag{3.42c}$$

where G is the Newtonian gravitational constant and $\mathscr{M}$ and m the active and passive gravitational masses, respectively. Then one obtains from (3.24), for the accuracy $\Delta\Gamma$ of the gravitational field $\Gamma^{\lambda}_{\mu\nu}$, the inequality

$$\Delta\Gamma \gtrsim \frac{\hbar}{cL_0^2 m}. \tag{3.43}$$

Assuming, following Peres and Rosen (1960) and Rosenfeld (1966), that the gravitational radius of the measurement body must be smaller than the length of the measurement body,

$$L_0 \gtrsim \frac{G\mathscr{M}}{c^2}, \tag{3.44}$$

one has for the Christoffel symbols

$$\Delta\Gamma L_0^3 \gtrsim \frac{\hbar G}{c^3}\frac{\mathscr{M}}{m} \tag{3.45}$$

((3.44) is equivalent to relation (3.16)). In the nonrelativistic approximation, the active gravitational mass $\mathscr{M}$ is, due to Newton's action—reaction law, equal to the passive gravitational mass m. Since, by virtue of the strong principle of equivalence, the same is valid in Einstein's GRT, one obtains generally

$$\Delta\Gamma L_0^3 \gtrsim \frac{\hbar G}{c^3}, \tag{3.46a}$$

$$\Delta g L_0^2 \gtrsim \frac{\hbar G}{c^3}. \tag{3.46b}$$

Accordingly, one sees that fundamental limitations on measurement occur which are associated with Planck's length

$$l_P = \left(\frac{\hbar G}{c^3}\right)^{1/2}. \tag{3.47}$$

The same result arises if, instead of (3.44), one uses the relation

$$L_0 \gtrsim \frac{Q^2}{Mc^2} \tag{3.48}$$

or

$$L_0 \gtrsim \frac{G\mathscr{M}^2}{Mc^2} \tag{3.49}$$

meaning that the field energy of the measurement body is smaller than its rest energy. This assumption results from the Bohr—Rosenfeld requirement that the measurement body is classical. With (3.49) one obtains from (3.43)

$$\Delta\Gamma L_0^3 \gtrsim \frac{\hbar G}{c^3}\,\frac{\mathscr{M}^2}{Mm}. \tag{3.50}$$

Considering that GRT satisfies the strong principle of equivalence, reading in its Newtonian limit,

$$m = \mathscr{M} = M, \tag{3.51}$$

we have again (3.46a) (see von Borzeszkowski and Treder, 1982b; Treder, 1975b).

For our discussion it is now essential to remark that all the measurement arguments given until now presuppose a sufficiently flat and rigid spacetime background in the domain of order L_0, so that classical rigid bodies can be moved without deformation. This was just one of the Bohr—Rosenfeld conditions on the measurement body. Assuming now a curved background somehow specified and a field which is to be measured, one must expect that the possibility of performing exact measurements is decreased (for a discussion of this point, see von Borzeszkowski and Treder, 1982b).

To estimate the restrictions on the measurements resulting from background curvature, let us consider the metric g near the point $x = 0$ in Riemannian coordinates. The uncertainties stemming from this then add to the uncertainties $\Delta g(0)$ discussed above. They amount to

$$\Delta g(x) = \Delta g(0) + \frac{\alpha}{|L^2|}\, x^2 \tag{3.52}$$

where $g(0)$ is the average value of g over a domain V of order L_0 measured under the assumption of a flat background, $g(x)$ is the average value of g taking the curvature into account, $1/L$ is a measure of the Riemannian background curvature, and α is a numerical constant greater than zero and of the order one. (L_0 is assumed to be much smaller than L, $L_0 \ll L$). From this it follows, with the aid of (3.50), that $\Delta g(x)$ satisfies the relation

$$\Delta g(x) \gtrsim \frac{l_P^2}{L^2} + \alpha\,|L|^{-2} L_0^2. \tag{3.53}$$

Comparing (3.46) with the flat-background relation (3.50) one sees that, for a curved background, there arises a larger uncertainty Δg. For $L \neq \infty$, upon minimization with respect to L_0, Δg reduces to

$$\Delta g(x) \gtrsim \frac{l_P}{|L|}\, \alpha^{1/2}. \tag{3.54}$$

This minimal value of Δg is approached for

$$L_0 \sim (1/\alpha)^{1/4} (L l_P)^{1/2}. \tag{3.55}$$

From (3.50) and (3.53) one may conclude that a measurement taking into account the action of a curved background leads to stronger and, in principle, more limitations on the measurement of fields than the usual quasi-Euclidean one. Of course, there arises the question of whether such an arrangement is in accordance with the requirements of measurement formulated by Bohr and Rosenfeld. Possibly a physical measurement always requires a suitable mechanism which compensates curvature effects. Indeed, an apparatus measuring curvature effects should be presupposed to be a uncurved rigid body according to which non-vanishing curvature is related. It is obvious that such a compensation mechanism may only be established if the curvature effect is well defined and *a priori* computable. This should be possible for a sufficiently static or even constant-curvature background. If we assume, however, that rapidly changing gravitational fields which produce a dynamic non-compensatable background are also measurable, then the measurements should be associated with the stronger limitations given by (3.53) (see Chapter 5).

The fact that an explicitly invariant approach does not change anything becomes especially evident from DeWitt (1964) and from the path integral discussion given above. In both cases our starting point, namely the Bohr—Rosenfeld uncertainty formula (3.25), is reproduced.

Quoting here the DeWitt paper in support of our arguments, a remark should be made in order to avoid misunderstandings. In DeWitt (1964) the conclusion is drawn from (3.25) that 'a single observable can always, in principle, be measured with arbitrary accuracy'. It is added that, if this assumption is invalid, the foundations of the quantum theory iteslf must be altered. However, the point we stress in this section is that for gravity, where (3.25) takes the form (3.46), there also exist limitations on the measurability of a single observable. Regarding this difference between gravitational and other (e.g., electromagnetic) fields, the second part of the DeWitt statement shows the consequences which (3.46) will have for the quantum theory of gravitation.

3.3. Ehrenfest's Theorems

In connection with gravitational wave experiments, limitations were discussed on the measurement of gravitational fields resulting from the quantum structure of the measurement device. It was, in particular, shown by Caves *et al.* (1980) that, considering a gravitational field detector as a quantum oscillator, the quantum dispersion of, for instance, the Weber detector is greater than the displacement induced by the gravitational field considered. On the other hand, it was mentioned by Dodonov *et al.* (1983) that, as a matter of principle, there cannot arise quantum limitations in measuring classical external forces. This was described as an almost trivial consequence of the Ehrenfest theorems because these theorems imply that quantum mechanics does not impose restrictions on the accuracy of the measurement of average values. Therefore, as is argued further, the classical force $f(t)$, i.e., the difference $f(t'') - f(t)$ of f at different time points t' and t'' should be measured with an unlimited accuracy via the Ehrenfest relation

$$f(t) = m \frac{\mathrm{d}^2}{\mathrm{d}t^2} \langle \mathbf{q} \rangle + \left\langle \frac{\partial u}{\partial q} \right\rangle, \tag{3.56}$$

where $u(\mathbf{q})$ is a potential, in particular an oscillator potential, and $\langle \ \rangle$ denotes the average value.

The latter argument notes the really trivial fact that, for the measurement of classical forces by *classical* measurement devices, there cannot arise quantum restrictions. One has, however, to take into account that all bodies, including measurement devices, have an atomistic structure. Considering them as classical bodies, one of course meets limits. Some of them could have a principle character. At any rate, this is true for the limitations given by Heisenberg's uncertainty relations. These relations also restrict the validity of the Ehrenfest theorem. Therefore, the above-given argument based on this theorem is nontrivial as far as it presupposes that, despite Heisenberg's uncertainty relations, a *classical* measurement can be realized with an, in principle, unlimited accuracy.

To clarify to what extent this presupposition is justified, we shall discuss quantum nondemolition measurements from the viewpoint of principles of measurement discussed above (we follow here von Borzeszkowski and Treder, 1983a). In our discussion we call the measurement of an observable a quantum nondemolition measurement if it is, in principle, possible to find a way to measure it, despite the quantum nature of the measuring device and do so with an arbitrarily high precision. Problems of such measurements were also investigated by Caves *et al.* (1978), Braginsky *et al.* (1977, 1978), Unruh (1978, 1979), and Gusev and Rudenko (1979).

To this end let us consider the character of the estimation for the minimum discoverable force. Following the authors cited above we assume as an 'antenna' a quantum harmonic oscillator described by the Hamiltonian operator

$$\mathbf{H} = \frac{1}{2m}\,\mathbf{p}^2 + \tfrac{1}{2}\,m\omega^2\mathbf{q}^2 - f(t)\mathbf{q}, \tag{3.57}$$

where $f(t)$ is an external classical force that is to be measured via its influence on the oscillator. The average values of the operator of the coordinate and the momentum change according to the equations of classical mechanics (Ehrenfest theorems)

$$\frac{\mathrm{d}}{\mathrm{d}t}\langle\mathbf{q}\rangle = \frac{1}{m}\langle\mathbf{p}\rangle, \tag{3.58}$$

$$\frac{\mathrm{d}}{\mathrm{d}t}\langle\mathbf{p}\rangle = -m\omega^2\langle\mathbf{q}^2\rangle + f(t) \tag{3.59}$$

(m is the mass, ω the frequency of the oscillator). The solutions to these equations are:

$$\langle\mathbf{q}(t)\rangle = q^0\cos\omega t + \frac{p^0}{m\omega}\sin\omega t + $$
$$+\frac{1}{m\omega}\int_0^t f(\tau)\sin[(t-\tau)\omega]\,\mathrm{d}\tau, \tag{3.60}$$

$$\langle\mathbf{p}(t)\rangle = -m\omega q^0\sin\omega t + p^0\cos\omega t + $$
$$+\int_0^t f(\tau)\cos[\omega(t-\tau)]\,\mathrm{d}\tau, \tag{3.61}$$

where $q^0 \equiv \langle \mathbf{q}(0) \rangle, p^0 \equiv \langle \mathbf{p}(0) \rangle$.

Considering the resonance case

$$f(t) = F_0 \sin(\omega t + \varphi),\ \varphi = \text{const.}, \tag{3.62}$$

and assuming $q^0 = p^0 = 0$ and $t \gtrsim \omega^{-1}$, one obtains for the maximum displacement of $\langle \mathbf{q} \rangle$:

$$\delta_q = \frac{F_0 t}{2m\omega}. \tag{3.63}$$

To measure a classical force $f(t)$ with an oscillator, one has of course to identify a displacement of the macroscopic oscillator coordinate $\langle \mathbf{q} \rangle$ as the action of the force under consideration. Therefore, δ_q must be greater than the square root $\sqrt{\sigma_q}$ of the dispersion $\sigma_q = \langle (\mathbf{q} - \langle \mathbf{q} \rangle)^2 \rangle$. The equations for the time evolution of the dispersion, resulting from the Schrödinger equation, and the Heisenberg uncertainty relation for σ_p and σ_q give

$$\sigma_q(t) \gtrsim \frac{\hbar}{2m\omega} |\sin \omega t|. \tag{3.64}$$

From (3.63) and (3.64) one obtains finally

$$\delta_q^2 \gtrsim \frac{h}{2m\omega} \tag{3.65}$$

or

$$F_0 \gtrsim \frac{1}{t} (2mh\omega)^{1/2} \tag{3.66}$$

(see, e.g., Dodonov *et al.*, 1983).

Considering a force measurement via an energy (or, in general, a quantum state) change, one obtains, instead of (3.66), the limits

$$F_0 \gtrsim \frac{1}{L_0} \left(\frac{\hbar}{m\omega} \right)^{1/2} \tag{3.67}$$

(L_0 is the linear dimension of the antenna) (see, e.g., Grishchuk and Polnarev, 1980).

Accordingly, one obtains limitations on the sensitivity. One sees, however, that the above-made considerations do not establish absolute limits. Indeed, except for $\hbar$, all the parameters occurring on the right-hand sides of (3.66) and (3.67) are wave- and antenna-dependent and not absolute constants. Therefore, one can overcome or defeat those limits, in principle, and it makes sense to look for a 'quantum nondemolition measurement'.

There is however a suspicious point in the discussion presented above. All arguments used have made no mention of the physical difference between gravitational and electromagnetic fields. In principle, all the arguments given above are also satisfied for electromagnetic field measurements. They do not try to take account of the difference between gravity and electromagnetism but work in an approximation where that difference is not effective.

The above-given measurement discussion based on the principle of equivalence makes the following point quite clear. In both classical and quantum gravity one has to assume a classical measurement body. The absolute limitations nevertheless arising result from Heisenberg's uncertainty relation together with some fundamental features of gravitational coupling, i.e., from laws on the interaction between field and test body. Therefore, in gravity one finds absolute limitations on L_0 measurements and, in general, on measurements of all entities having the dimension length, time or mass. Indeed, due to the occurrence of the three universal constants $\hbar$, c, G, one may build the Planckian length, time and mass units

$$l_P = \left(\frac{\hbar G}{c^3}\right)^{1/2}, \; t_P = \left(\frac{\hbar G}{c^5}\right)^{1/2}, \; m_P = \left(\frac{\hbar c}{G}\right)^{1/2}. \tag{3.68}$$

They define the dimension of the smallest classical test body and, simultaneously, the biggest elementary particle in a Grand Unified Theory incorporating general relativity (for details, see Section 6.2). They define accordingly the border between a measurement body and the system whose properties are measured. This border causes the limitations on measurement which one cannot evade in a full ($\hbar$, c, G) theory (see Chapter 1). It results (1) from the limits on action measurements characterized by $\hbar$, (2) from the limits on synchronous measurements characterized by the velocity of light in vacuum c, and (3) from the universal coupling of gravitation and matter characterized by G.

The considerations of the authors cited above aimed at the establishment of 'quantum nondemolition measurements' which overcome limitations resulting from the quantum structure of gravitational antennas, may be interpreted as follows. They discuss physical entities and approximations to the full ($\hbar$, c, G) theory so that absolute limitations do not appear. If one considers, e.g., a quantum oscillator in the non-relativistic approximation without taking into consideration the universal character of gravity, then there do not arise absolute limits of length etc. Then only $\hbar$, and not $\hbar$, c and G, restricts the sensitivity of antennas.

Furthermore, considering quantum electrodynamics containing the constants $\hbar$ and c, one obtains a charge $Q = (\hbar c)^{1/2}$ and the corresponding limitation on the charge of the test body, $Q^2 \gtrsim hc$. To obtain, however, a universal mass or length and corresponding limitations on the test body one has yet to introduce a new universal constant relating charge and mass given by the universal constants $\hbar$, c, G. Without a universal coupling constant leading from special relativity to general relativity there is no relation between mass and charge; each complete theory contains all three Planck units; otherwise a theory is not physically interpretable. As long as one considers gravitational theory only without any explicit incorporation of G, one moves again within the frame of an approximation to the full $(\hbar, c, G)$ theory which epistemologically does not differ from electrodynamics. In reality, this means gravitation without gravitation.

One finds a similar situation in a $(\hbar, G)$ approximation considered in Chapter 2. Indeed, discussing physical processes describable by quantum mechanics and Newton's gravitational theory, one can always find quantities on which, in principle, no measurement limitations are imposed.

Measurements based on $(\hbar)$, $(\hbar, c)$ or $(\hbar, G)$ approximations fail, of course, to measure the effects of the full theory. Otherwise, measurements referring to the full theory suffer from absolute limitations (see also Chapter 6.2).

The Bohr—Rosenfeld proper measurements refer to a local Riemannian coordinate system, where the measurement is performed at the center P_0 of this system. In these coordinates, one has

$$(g_{00})_{P_0} = 1, \qquad (g_{ik})_{P_0} = -\delta_{ik}, \qquad (\Gamma^{\lambda}_{\mu\nu})_{P_0} = 0, (i, k = 1, 2, 3)$$

so that for an optimal measurement, where

$$\Delta x \lesssim x$$

and

$$\Delta g_{00} \leqslant g_{00} < 1,$$

the relation (3.46b) holds. A measurement made in any other but the proper system is of a lower accuracy. Indeed, due to the Lorentz contradiction, one has (cf. Appendix B)

$$\Delta g_{00}(\Delta x)^2 \gtrsim \hbar G/g_{00}c^3 \tag{3.69}$$

Accordingly, for $g_{00} < 1$ the limit of $\Delta g_{00} \cdot (\Delta x)^2$ increases. Therefore, any uncertainty relation resulting from (3.69) does not have a fundamental meaning. It is only an expression of the fact that one may measure worse than said by (3.46b). This is particularly true for the relation (cf. Unruh, 1984),

$$\Delta G_{00}\Delta g_{00}(\Delta x)^4 \gtrsim \hbar G/c^3$$

following, via (3.69), from $\Delta G_{00} \sim g_{00}/\Delta x$.

Chapter 4

Mathematical Descriptions of Quantum Gravity

4.1. Heisenberg—Euler—Kockel Approximation

The physical meaning of the conditions (3.46) imposed on gravitational field measurements becomes more obvious if one considers the mathematical description of the interaction between the field one intends to measure and the measurement body acting otherwise as a source of the field. Such considerations generalize our remarks related to formula (3.17) to the effective (phenomenological) equations of quantum gravity.

For this purpose, let us remember the well-known fact that the interaction of electromagnetic fields with Dirac's electron—positron vacuum leads to a self-interaction of those fields. Indeed, two photons with momenta $\mathbf{k}_1$ and $\mathbf{k}_2$, whose energy is too small to cause pair creation, will create two virtual pairs leading to two new photons with momenta $\mathbf{k}_3$ and $\mathbf{k}_4$. This process results in the two incident photons, $\mathbf{k}_1$ and $\mathbf{k}_2$, scattering each other and gives rise to two outgoing photons $\mathbf{k}_3$ and $\mathbf{k}_4$. This quantum effect cannot be described, of course, by the usual linear Maxwell equations, where the sum of two solutions is again a solution. For the nonlinearity resulting from coupling electromagnetic fields to Dirac matter, see Chapter 1. To describe the quantum effect, one has to discuss nonlinear modifications of the Larmor Lagrangian

$$\Lambda_0 = \tfrac{1}{4} F^{\mu\nu} F_{\mu\nu} \tag{4.1}$$

Such a modification was attempted by Heisenberg, Euler and Kockel (Euler and Kockel, 1935; Euler, 1936; Heisenberg and Euler, 1936) for the limit of low photon energy ($\hbar\omega \ll mc^2$).

Following this procedure to describe quantum field effects, we consider a field F which changes so softly that no real massive particle pairs are created. Then we have to discuss the pure field cases, and the corrections to the classical Lagrangian Λ_0, symbolically written as

$$\Lambda_0 = \tfrac{1}{4} F^2 \tag{4.2}$$

must be formed by a term of fourth order in F:

$$\Lambda_1 \sim \left(\frac{L^2}{Q}\right)^2 F^4. \tag{4.3}$$

F is assumed to be slowly changing. The correction depends thus only on the field F and not on its time and spatial derivatives. Furthermore, we do not consider here corrections of higher order than F^4.

Further constants, making (4.3) an equation, are of the order 1 if we put $Q^2 = \hbar c$; Q/L^2 determines the order of magnitude of the 'Dirac vacuum field' causing the nonlinear interaction.

Because we are discussing perturbations of measurement bodies, we have to assume Q/L^2 to be the field of those bodies. To guarantee now that the correction Λ_1 is really resulting from the action of a measurement body (and not from a usual source of quantum electrodynamics), one has to add again the Bohr—Rosenfeld conditions (3.15) and (3.16) on the source term; i.e., we assume

$$Q^2 \gtrsim \hbar c, \tag{4.4a}$$

$$L_0 > \frac{Q^2}{Mc^2}\,. \tag{4.4b}$$

In particular, they prevent us from repeating the mistake made by Landau and Peierls (see Section 3).

In the gravitational case, one obtains from (4.4) Rosenfeld's relations (Rosenfeld, 1966)

$$Gm^2 = G\mathcal{M}^2 \gtrsim \hbar c \tag{4.5a}$$

and

$$L_0 > \frac{G\mathcal{M}}{c^2}, \tag{4.5b}$$

where $G\mathcal{M}/c^2$ denotes the gravitational radius of the measurement body. For $L_0 \lesssim G\mathcal{M}/c^2$, the body degenerates into a black hole, i.e., to an information hole. Relations (4.4) and (4.5) optimize the interaction between the exterior field, F, and the measurement body, so that the action of this body perturbs the field (and its measurement) minimally. The question is whether this coupling is sufficiently weak that the quantum effects of the field F are not drowned in the sea of quantum perturbations caused by the measurement body. Technically speaking, one has to prove the compatibility of our conditions (4.4) or (4.5) with the Heisenberg—Euler—Kockel approximation, i.e., especially with the requirement that Λ_1 be a second-order correction to the first-order Lagrangian Λ_0. Otherwise, one could not realize any measurement equipment for measuring quantum effects of F. The condition

$$\left(\frac{L^2}{Q}\right)^2 F^2 < 1 \tag{4.6}$$

is thus at least necessary to be satisfied for the measurability of quantum effects.

In quantum electrodynamics, where (4.3) takes a Born—Infeld form (Born, 1934; Born and Infeld, 1934)

$$\Lambda = \Lambda_0 + \Lambda_1 = \tfrac{1}{4} F^{\mu\nu} F_{\mu\nu} + \\ + \left(\frac{L^2}{Q}\right)^2 [\alpha(F^{\mu\nu}F_{\mu\nu})^2 + \beta(F^{\mu\nu}F^*_{\mu\nu})^2] \tag{4.7}$$

(α and β are numerical constants of order of magnitude 1). Accordingly the following order-of-magnitude relation must hold:

$$\left(\frac{L^2}{Q}\right)^2 (F^{\mu\nu}F_{\mu\nu})^2 \sim \left(\frac{L^2}{Q}\right)^2 (F^{\mu\nu}F^*_{\mu\nu})^2 \sim \\ \sim \left(\frac{L^2}{Q}\right)^2 F^4 < F^2 \sim |F^{\mu\nu}F_{\mu\nu}|. \tag{4.8}$$

This requirement can be satisfied in a large region of field strength values. Only for $Q = (\hbar c)^{1/2}$, i.e., for the worst measurement, does this approximation break down.

Turning to the gravitational case, we have to consider the Einstein—Hilbert Lagrangian

$$\Lambda_0 = \sqrt{-g}\, g^{\mu\nu} R_{\mu\nu}, \tag{4.9}$$

where $R_{\mu\nu}$ is the Ricci tensor formed from the metric $g_{\mu\nu}$ and its first and second derivatives and g denotes the determinant of $g_{\mu\nu}$. Looking for quantum corrections in the spirit of the Heisenberg—Euler—Kockel approximation, one finds formally a similar situation as in electrodynamics.

Indeed, in Maxwell's electrodynamics there exists only the one Lorentz scalar (4.1) leading to linear field equations. Supplementarily, there are the two quadratic Lorentz invariants,

$$(F^{\mu\nu}F_{\mu\nu})^2 \text{ and } (F^{\mu\nu}F^*_{\mu\nu})^2. \tag{4.10}$$

(The term $F^{\mu\nu}F_{\mu\nu}F^{*\alpha\beta}F^*_{\alpha\beta}$ does not provide anything new.) One finds a similar situation in Einstein's GRT. The Ricci scalar R is the only invariant establishing second-order derivative equations for $g_{\mu\nu}$. The corrections which describe the quantum-physical influence of the measurement body on gravitational fields here are as follows

$$\sim \left(\frac{L^2c^2}{G\mathcal{M}}\right)^2 \alpha R_{\mu\nu}R^{\mu\nu} \tag{4.11}$$

and

$$\sim \left(\frac{L^2c^2}{G\mathcal{M}}\right)^2 \beta R^2. \tag{4.12}$$

(The term $\sim R_{\alpha\beta\gamma\delta}R^{\alpha\beta\gamma\delta}$ may be neglected because, in the four-dimensional case, it can be expressed by a linear combination of (4.11) and (4.12).)

Accordingly, we have to consider, instead of the Einstein—Hilbert Lagrangian (4.9), the Einstein—Eddington—Lanczos Lagrangian

$$\Lambda = \sqrt{-g}\left[R + \left(\frac{L^2c^2}{G\mathcal{M}}\right)^2(\alpha R_{\mu\nu}R^{\mu\nu} + \beta R^2)\right], \tag{4.13}$$

where α and β are again numerical constants of the order of magnitude 1 if $\sqrt{G\mathcal{M}}$ is measured in units of $\sqrt{\hbar c}$.

To satisfy (4.6) one has then to demand the order-of-magnitude relation

$$1 > \left(\frac{L^2c^2}{G\mathcal{M}}\right)^2 (R_{\mu\nu}R^{\mu\nu})^{1/2} \sim \left(\frac{L^2c^2}{G\mathcal{M}}\right)^2 R \sim \left(\frac{L^2c^2}{G\mathcal{M}}\right)^2 \Gamma^2. \tag{4.14}$$

In view of (4.5) and the principle of equivalence (3.51), one finds from (4.14)

$$1 > \frac{L^4c^4}{G^2\mathcal{M}^2}\Gamma^2 \gtrsim \left(\frac{G\mathcal{M}}{c^2}\right)^2 \Gamma^2 \gtrsim \frac{\hbar G}{c^3}\Gamma^2. \tag{4.15}$$

It says that the Heisenberg—Euler—Kockel approach, describing together with (4.5) the interaction between measurement body and field Γ, works only for curvatures that are weaker than

$$R_{\rm P} \sim \Gamma_{\rm P}^2 \sim \frac{1}{l_{\rm P}^2}, \tag{4.16}$$

where $l_{\rm P} = (\hbar G/c^3)^{1/2}$ again denotes Planck's length. This reproduces our result derived in Chapter 3 that there arise principle limitations on Γ (and accordingly on R) measurements. Below the length $l_{\rm P}$, quantum effects of the measurement body destroy the measurability of the Γ field.

In quantum electrodynamics the constants L^2/Q, α and β in the Born—Infeld-type Lagrangian (4.7) depend on the structure of the measurement body. This is evident (i) from the calculations of Euler (1936) for Dirac matter, considering terms corresponding to the diagram shown in Figure 4.1.

(the internal line is that of spin-$\frac{1}{2}$ fermions) and (ii) from the calculations of Schwinger (1951) for scalar matter shown in Figure 4.2 (the internal line is that of spin-0 bosons).

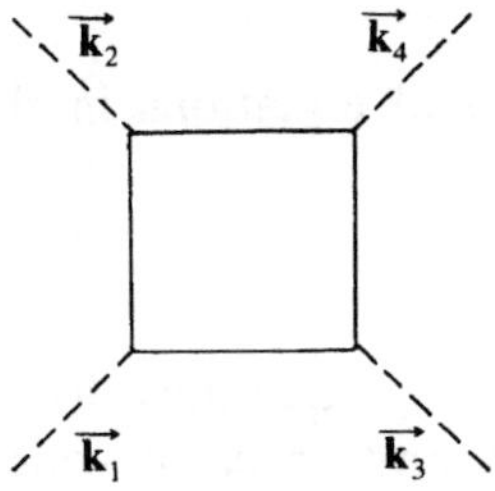

Fig. 4.1.

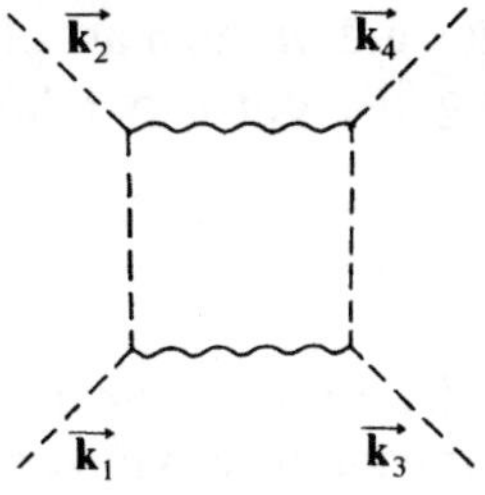

Fig. 4.2.

In GRT, the constants $L^2/G\mathcal{M}$, α and β in (4.13) have, in contrast to quantum electrodynamics, a universal character. α and β can be calculated so that

$$\frac{G\mathcal{M}}{L^2} = \frac{c^4}{G} \frac{2\pi}{(\hbar c)^{1/2}} \tag{4.17}$$

is a universal constant.

Maxwell's equations are linear field equations of first order in the derivatives of $F_{\mu\nu}$ and, accordingly, of second order in the derivatives of the vector potential A_μ. The electromagnetic equations of the Born—Infeld type resulting from (4.7) are again of the same order but nonlinear. According to the structure of the invariants R^2 and $R_{\mu\nu}R^{\mu\nu}$, the Einstein—Eddington—Lanczos field equations stemming from (4.13) are however nonlinear equations of fourth order. The influence of the measurement body is therefore to be taken into account already for 'weak fields' Γ.

The gravitational fourth-order equations are reminiscent of the 'electrodynamical higher-derivative equations' proposed by Bopp (1940) and Podolsky (1941). Those equations result from the Lagrangian

$$L_{\mathrm{BP}} = \tfrac{1}{4}F^{\mu\nu}F_{\mu\nu} + \frac{\alpha_0}{l^2} F_{\mu\nu,\,\alpha}F^{\mu\nu,\,\alpha} \tag{4.18}$$

(α_0 is a dimensionless constant of order of magnitude 1 and l is a length parameter). As is well known, the Bopp—Podolsky equations describe an electromagnetic field to which not only restmassless photons but also heavy particles (a second sort of photons or W bosons) are attributed (see, e.g., Treder, 1974). The rest mass of the heavy particles is given by their Compton wavelength l appearing in (4.18),

$$l = \hbar/Mc. \tag{4.19}$$

Bopp and Podolsky specified l as the classical electron radius such that M is given by

$$M = \frac{\hbar c}{e^2} m \tag{4.20}$$

(e is the charge and m the mass of electrons).

Considering the corresponding gravitational equations in the Eddington case, where

$$\alpha = -2\beta \tag{4.21}$$

then one finds, in the linear approximation, equations for a symmetric second-rank tensor in a Minkowski background space to which, beside the zero rest mass gravitons, heavy gravitons with mass

$$M = \left(\frac{\hbar G}{c^3}\right)^{1/2} \frac{c^2}{G} \tag{4.22}$$

are attributed (for details see Section 6.1). The appearance of massive gravitons is an expression of the influence of the measurement body on gravitational fields. It would vanish when GM/c^2 became infinitely large. Due to (4.22), this means that the rest mass of heavy gravitons had to be infinitely large to suppress the perturbation of the measurement body.

Now it is quite clear that it makes no sense to aim at really infinitely heavy particles. Infinitely heavy gravitons providing a negligible influence of the measurement body are, for instance, equivalent to a vanishing coupling between gravity and nongravitational matter, and this excludes, of course, any possibility of measurement. The problem arising in quantum gravity is, however, that the rest mass of heavy gravitons is universally given by the mass of planckions. Therefore, the accuracy of measurements cannot be unlimitedly improved. Planckions are optimal measurement bodies.

In other words, in GRT gravitodynamics the description of quantum coupling between matter and gravity, performed in the sense of the Heisenberg—Euler—Kockel approximation by Born—Infeld type corrections to $\Lambda_0 = \sqrt{-g}\, R$, is — due to the strong equivalence principle — equivalent to fourth-order corrections providing heavy gravitons. Just this equivalence causes limitations on measurement. In Chapters 5 and 6 we shall discuss the meaning of these limitations for an experimental verification of quantum gravity.

Before going into particulars, let us consider here a general argument signalling the problems which arise from the limitations on measurements. For this purpose, we follow Einstein (1911), who formulated the following essential test for the profundity of physical relations (i.e., of physical theories and their relations). The dimension—analytical expressions which can be formed of the constants arising in a theory under consideration should only differ by a factor of order of magnitude 1 from the actually measurable quantities of the same dimension.

Considering quantum gravity with its three universal constants $\hbar$, c, G and accepting Einstein's criterion (under the presupposition that quantum gravity is a profound theory), one finds that all measurable quantities of quantum gravity have the order of Planck's units (see Chapter 3.3). Otherwise the limitations on measurements are of this order, as we have mentioned repeatedly. The conclusion one must draw from this is, of course, that one has to expect difficulties in measuring quantum effects of Einstein's GRT.

4.2. On Gauge Fixing in Quantum Gravity

Whether quantum effects of gravity are measurable or not, quantization of gravitational fields is, as stressed repeatedly above, unavoidable for a consistent description of the interaction of gravitational and quantized matter fields. There arise, however, problems in the application of the usual quantization schemes to gravity, because such procedures have to evade modifications which destroy the classical GRT one wants to quantize (see our remarks in Chapter 1). Before continuing the measurement discussion, we shall consider here one aspect of this problem by discussing gauge invariance and its breaking in the covariant quantization approach (see Bleyer and Borzeszkowski, 1984).

The covariant quantization of the gravitational field uses the functional integral formulation of quantum field theory (DeWitt, 1964; Faddeev and Popov, 1973). This formalism starts from the action integral for the gravitational field

$$I_{\mathrm{grav}} = \int \mathrm{d}^4x \sqrt{-\mathrm{g}}\, L_{\mathrm{grav}}[g_{\mu\nu}] \tag{4.23}$$

and calculates the transition amplitudes from the functional integral

$$Z = \int \mathrm{d}[g_{\mu\nu}] \exp(\mathrm{i} I_{\mathrm{grav}}[g_{\mu\nu}]), \tag{4.24}$$

where $g_{\mu\nu}$ denotes the Riemannian metric of the spacetime manifold. The Lagrangian density is given by the Ricci scalar $L_{\mathrm{grav}} = R$. Following the Faddeev—Popov formalism of quantized gauge fields (Faddeev and Slavnov, 1978), the functional integral is to be carried out over all nongauge equivalent classical field configurations. This is realized by choosing a section of the gauge group which intersects all orbits of the group one times. The orbits of the gauge group are parametrized by a gauge fixing term which is added to the classical action together with a compensating Faddeev—Popov determinant. For the application of this formalism to the gravitational field a suitable gauge fixing term is given by (Duff, 1975)

$$L_{gf} = \frac{\eta_{\mu\nu}}{g} (\sqrt{-g}\, g^{\mu\lambda})_{,\lambda} (\sqrt{-g}\, g^{\nu\tau})_{,\tau}, \tag{4.25}$$

where $\eta_{\mu\nu}$ denotes the metric of the Minkowski spacetime. This gives Einstein's GRT in the de Donder gauge. Generalizations of this gauge fixing introduce an arbitrary fixed reference metric $g^{\mathrm{R}}_{\mu\nu}$ replacing the flat metric so that L_{gf} is a scalar with respect to coordinate transformations constructed from both metrics $g^{\mathrm{R}}_{\mu\nu}$ and $g_{\mu\nu}$. This procedure is based on the following understanding of gauge invariance and gauge fixing in gravitation.

Due to the validity of the strong principle of equivalence in GRT, the gravitational field is described by the Riemannian metric $g_{\mu\nu}$ alone. The Einstein group of coordinate transformations plays, therefore, a double role. As the group of coordinate transformations it ensures the free choice of coordinate systems required by the covariance principle. Otherwise it is the gauge group of the tensor $g_{\mu\nu}$. If a reference metric $g^{\mathrm{R}}_{\mu\nu}$, globally fixed, is incorporated, the two actions of the Einstein group can be separated. The reference metric $g^{\mathrm{R}}_{\mu\nu}$ and the metric $g_{\mu\nu}$ transform both as tensors with respect to coordinate transformations. This means that, in order to ensure coordinate covariance, action integrals have to be scalars constructed from $g^{\mathrm{R}}_{\mu\nu}$ and $g_{\mu\nu}$.

As a gauge group the Einstein group acts on $g_{\mu\nu}$ alone, $g^{\mathrm{R}}_{\mu\nu}$ remains fixed. Therefore, a gauge transformation changes $g_{\mu\nu}$ with respect to $g^{\mathrm{R}}_{\mu\nu}$. Gauge invariance in gravitation consists in the free relative orientation of the two metrics $g_{\mu\nu}$ and $g^{\mathrm{R}}_{\mu\nu}$. According to this statement, gauge fixing is realized by the addition to the action integral of an arbitrary gauge-fixing term of the structure*

$$L_{gf} = L_{gf}[g^{\mathrm{R}}_{\mu\nu}, g_{\mu\nu}] \tag{4.26}$$

specifying $g_{\mu\nu}$ with respect to $g^{\mathrm{R}}_{\mu\nu}$. Examples of this kind are given by Adler (1982). The scheme must be completed by the proof that the chosen gauge is an allowed one (Faddeev and Slavnov, 1978).

Usually the background field method (DeWitt, 1964) is used, according to which the metric is decomposed into a classical part and quantum fluctuations

$$g_{\mu\nu} = g^{\mathrm{B}}_{\mu\nu} + h_{\mu\nu}. \tag{4.27}$$

In this case the reference metric $g^{\mathrm{R}}_{\mu\nu}$ can be identified with the background metric $g^{\mathrm{B}}_{\mu\nu}$. This avoids the appearance of external structures artificial to the framework of GRT. Without the identification of the two metrices $g^{\mathrm{R}}_{\mu\nu}$ and

* This means that L has a Rosen—Kohler-type structure.

$g^{\mathrm{B}}_{\mu\nu}$ we have, instead of gauged GRT, a bimetric theory showing classical consequences apart from GRT.

However, the identification of $g^{\mathrm{R}}_{\mu\nu}$ and $g^{\mathrm{B}}_{\mu\nu}$ is not allowed for arbitrary L_{gf}, contrary to the Faddeev—Popov formalism, where arbitrary gauge fixing terms of the structure (4.26) are possible. This means that not all L_{gf} are gauge-fixing terms for GRT, but terms producing bimetric modifications of GRT.

In Einstein field equations, the addition of L_{gf} leads to an additional term

$$T^{\mathrm{R}}_{\mu\nu} = \frac{\delta L_{gf}}{\delta g^{\mu\nu}}. \tag{4.28}$$

If the reference metric is treated as an additional matter field, we have the field equations

$$\frac{\delta L_{gf}}{\delta g^{\mathrm{R}}_{\mu\nu}} = 0 \tag{4.29}$$

and from the invariance property of L_{gf} with respect to coordinate transformations, we have

$$\frac{\delta L_{gf}}{\delta g_{\mu\nu}}\, \delta g_{\mu\nu} + \frac{\delta L_{gf}}{\delta g^{\mathrm{R}}_{\mu\nu}}\, \delta g^{\mathrm{R}}_{\mu\nu} = 0. \tag{4.30}$$

With the help of (4.29), this gives the conservation law

$$T^{\mathrm{R}\nu}_{\mu;\,\nu} = 0. \tag{4.31}$$

The same condition follows from the gravitational field equation using Bianchi identities (see below). Equation (4.31) leads to four conditions fixing the relative orientation of the two metrics $g_{\mu\nu}$ and $g^{\mathrm{R}}_{\mu\nu}$. We really need four equations for the metric in order to fix the gauge, because the gauge transformation law contains four arbitrary functions.

As a special example we consider a L_{gf} which contains the reference metric $g^{\mathrm{R}}_{\mu\nu}$ only in the combination of the Ricci tensor $R^{\mathrm{R}}_{\mu\nu}$. From the great variety of possible terms, we use

$$L_{gf} = -g^{\mu\nu} R^{\mathrm{R}}_{\mu\nu}. \tag{4.32}$$

The combination of this L_{gf} with the Einstein—Hilbert Lagrangian leads to Rosen's bimetric theory with arbitrary background (Rosen, 1979). In order to show this we start for I_{cl} from the expression

$$I_{cl} = I_{\mathrm{grav}} + I_{gf} = \int K \sqrt{-g}\, \mathrm{d}\Omega$$

$$= \int \gamma K \sqrt{-g^{\mathrm{R}}}\, \mathrm{d}\Omega. \tag{4.33}$$

Here we use the notations

$$K = K_{\mu\nu}g^{\mu\nu} = g^{\mu\nu}(R_{\mu\nu} - R^{\mathrm{R}}_{\mu\nu}) \tag{4.34}$$

and

$$\gamma = (g/g^{\mathrm{R}})^{1/2}. \tag{4.35}$$

Equation (4.34) can be written in the form

$$K = g^{\mu\nu}(\Delta^{\lambda}_{\mu\nu|\lambda} - \Delta^{\lambda}_{\mu\lambda|\nu} + \Delta^{\lambda}_{\mu\nu}\Delta^{\tau}_{\lambda\tau} - \Delta^{\lambda}_{\mu\tau}\Delta^{\tau}_{\nu\lambda}), \tag{4.36}$$

where $\Delta^{\lambda}_{\mu\nu}$ denotes the difference between the Christoffel symbols $\left\{{\lambda \atop \mu\nu}\right\}$ constructed from $g_{\mu\nu}$ and the Christoffel symbols $\Gamma^{\lambda}_{\mu\nu}$ constructed from $g^{\mathrm{R}}_{\mu\nu}$. "|" means the covariant differentiation with respect to the reference metric $g^{\mathrm{R}}_{\mu\nu}$. In the presence of matter field, we have the classical action

$$I_{cl} = \frac{c^3}{16\pi k}\int K\sqrt{-g}\, \mathrm{d}\Omega + \int L_{\mathrm{M}}\sqrt{-g}\, \mathrm{d}\Omega. \tag{4.37}$$

From this action we derive the field equations

$$G_{\mu\nu} = \frac{8\pi k}{c^4} T^{\mathrm{M}}_{\mu\nu} + S_{\mu\nu} \tag{4.38}$$

with

$$S_{\mu\nu} = R^{\mathrm{R}}_{\mu\nu} - \tfrac{1}{2} g_{\mu\nu}g^{\lambda\tau}R^{\mathrm{R}}_{\lambda\tau} = T^{\mathrm{R}}_{\mu\nu}. \tag{4.39}$$

Indeed, we arrived at Einstein's equations with an additional source term. Just this property of L_{gf} leads to difficulties in the interpretation of L_{gf} as a gauge breaking term. From the field equations we derive the following covariant divergency

$$G^{\nu}_{\mu;\,\nu} = \kappa T^{\mathrm{M}\nu}_{\mu;\,\nu} + T^{\mathrm{R}\nu}_{\mu;\,\nu} = 0. \tag{4.40}$$

If we demand that the dynamical equations

$$T^{\mathrm{M}\nu}_{\mu;\,\nu} = 0 \tag{4.41}$$

of nongravitational matter be fulfilled and use the Bianchi identities, we get the conditions (4.31)

$$T^{\mathrm{R}\nu}_{\mu;\,\nu} = 0 \tag{4.42}$$

fixing the relation between the two metrics $g_{\mu\nu}$ and $g^{\mathrm{R}}_{\mu\nu}$. Only in the case of a flat reference metric $g^{\mathrm{R}}_{\mu\nu}$ are the conditions (4.42) fulfilled identically. This is, therefore, the only gauge invariant bimetric theory because the orientation of the metric $g_{\mu\nu}$ with respect to the background is arbitrary. This theory is known as the Einstein—Rosen theory.

In the model considered here, different special gauge conditions of the Riemannian metric $g_{\mu\nu}$ can be realized. Using the conditions (4.42), we calculate which reference metrics have to be chosen in order to fix this special gauge.

Let us consider the following choice. In order to have spin-2 gravitons, small perturbations $h_{\mu\nu}$ have to fulfil the Hilbert condition in flat space

$$(h^{\nu}_{\mu} - \tfrac{1}{2}\,\delta^{\nu}_{\mu} h)_{,\mu} = 0. \tag{4.43}$$

There are different possibilities for writing these conditions covariantly with respect to the reference metric $g^{\mathrm{R}}_{\mu\nu}$. For example we can use

$$(g^{\mu\nu} - \tfrac{1}{2}\,g^{\mathrm{R}\mu\nu} g^{\mathrm{R}}_{\lambda\tau} g^{\lambda\tau})_{|\nu} = 0 \tag{4.44}$$

or the covariant version of the de Donder condition

$$(\sqrt{-g}\, g^{\mu\nu})_{|\nu} = 0. \tag{4.45}$$

We ask now for the possible reference metrics, for which the equation (4.42) for the $g^{\mathrm{R}}_{\mu\nu}$ leads to condition (4.45) for the metric $g_{\mu\nu}$. Equation (4.42) can be written in the form

$$R^{\mathrm{R}}_{\mu\nu;\lambda} g^{\nu\lambda} - \tfrac{1}{2}\,\delta^{\lambda}_{\mu} R^{\mathrm{R}}_{\nu\tau;\lambda} g^{\nu\tau} = 0 \tag{4.46}$$

and the conditions (4.45) read

$$(g^{\mu\nu}_{|\nu} - \tfrac{1}{2}\,g^{\lambda\tau}_{|\nu} g_{\lambda\tau} g^{\mu\nu})\, g^{\mathrm{R}}_{\mu\sigma} = 0. \tag{4.47}$$

The calculation of the covariant derivatives gives

$$\begin{aligned}(g^{\mu\nu}_{|\nu} - \tfrac{1}{2}\,g^{\lambda\tau}_{|\nu} g_{\lambda\tau} g^{\mu\nu})\, g^{\mathrm{R}}_{\mu\sigma} \\ = g^{\mathrm{R}}_{\sigma\mu;\lambda} g^{\mu\lambda} - \tfrac{1}{2}\,\delta^{\lambda}_{\sigma} g^{\mathrm{R}}_{\mu\nu;\lambda} g^{\mu\nu},\end{aligned} \tag{4.48}$$

Therefore (4.45) is equivalent to

$$g^{\mathrm{R}}_{\sigma\mu;\lambda} g^{\mu\lambda} - \tfrac{1}{2}\,\delta^{\lambda}_{\sigma} g^{\mathrm{R}}_{\mu\nu;\lambda} g^{\mu\nu} = 0. \tag{4.49}$$

Then the coincidence of (4.42) and (4.45) is given for reference metrics with constant curvature

$$R^{\mathrm{R}}_{\mu\nu} = \text{const} \times g^{\mathrm{R}}_{\mu\nu}. \tag{4.50}$$

These are the metrics considered by Rosen (1979) as global cosmic background metrics.

In order to determine whether (4.28) can be used as a gauge-fixing term, we return to the background-field method and separate, according to (4.27), the classical background metric $g^{\mathrm{B}}_{\mu\nu}$ from $g_{\mu\nu}$ and discuss the possibility of an identification of $g^{\mathrm{B}}_{\mu\nu}$ and $g^{\mathrm{R}}_{\mu\nu}$.*

* Since small perturbations of the gravitational field propagate along the bicharacteristic defined by $g^{\mathrm{B}}_{\mu\nu}$, for gravitational perturbations (gravitational waves) as well as for nongravitational matter, the spacetime background is the physical metric. Therefore, the reference metric can be given a physical meaning of the background only if the reference metric can be identified with the spacetime background.

We consider this problem for the vacuum field equations of Rosen's model

$$R_{\mu}^{\nu} = R_{\mu}^{R\nu} \tag{4.51}$$

using the so-called high-frequency approximation, which allows us to study back-reaction (Brill and Hartle, 1964; Isaacson, 1967).

To this end, we assume the metric $g_{\mu\nu}$ to be represented by the ansatz

$$g_{\mu\nu} = g_{\mu\nu}^{B} + \varepsilon h_{\mu\nu}, \tag{4.52}$$

where

$$g_{\mu\nu}^{B} \sim h_{\mu\nu} = O(1), \tag{4.53}$$

$$\partial g_{\mu\nu}^{B} \sim \frac{g_{\mu\nu}^{B}}{L}, \ \partial h_{\mu\nu} \sim \frac{h_{\mu\nu}}{\lambda}, \tag{4.54}$$

$$\varepsilon \ll 1, \lambda \ll L. \tag{4.55}$$

The background metric $g_{\mu\nu}^{B}$ for gravitational perturbations is a slowly varying function of x^{μ} and is generated by some averaging procedure over the high-frequency perturbations $h_{\mu\nu}$. L and λ are characteristic lengths over which the background and the short-wave field $h_{\mu\nu}$ change significantly; ε is a smallness parameter. To complete this scheme, we assume furthermore that the characteristic length over which the fix background $g_{\mu\nu}^{R}$ changes is denoted by l:

$$\partial g_{\mu\nu}^{R} \sim g_{\mu\nu}^{R}/l. \tag{4.56}$$

Inserting the ansatz (4.52) into the Ricci tensor $R_{\mu\nu}$, we obtain

$$R_{\mu\nu} = R_{\mu\nu}^{(0)} + \varepsilon R_{\mu\nu}^{(1)} + \varepsilon^{2} R_{\mu\nu}^{(2)} + \dots, \tag{4.57}$$

where the terms $R_{\mu\nu}^{(0)}$, $\varepsilon R_{\mu\nu}^{(1)}$, $\varepsilon^{2} R_{\mu\nu}^{(2)}$ are of the following orders of magnitude

$$R_{\mu\nu}^{(0)} \sim \frac{1}{L^{2}}, \ \varepsilon R_{\mu\nu}^{(1)} \sim \frac{\varepsilon}{\lambda^{2}}, \ \varepsilon R_{\mu\nu}^{(2)} \sim \frac{\varepsilon^{2}}{\lambda^{2}}. \tag{4.58}$$

$R_{\mu\nu}^{R}$ is of the order of magnitude l^{-2}. In this expansion the only linear term in ε is $R_{\mu\nu}^{R}$. We therefore put

$$\varepsilon R_{\mu\nu}^{(1)} = 0. \tag{4.59}$$

The next approximation divides $R_{\mu\nu}$ in a part averaged over some wavelength λ and small nonlinear perturbations describing self-interaction effects of the wave. We have at this stage

$$R_{\mu\nu} = R_{\mu\nu}^{(0)} + \langle R_{\mu\nu}^{(2)} \rangle. \tag{4.60}$$

If we assume $l \sim L$ we get the field equations

$$R_{\mu\nu}^{(0)} + \langle R_{\mu\nu}^{(2)} \rangle = R_{\mu\nu}^{R}. \tag{4.61}$$

According to Brill and Hartle (1964), the expression

$$-\langle R^{(2)}_{\mu\nu} - \tfrac{1}{2} g^{\mathrm{B}}_{\mu\nu} g^{\mathrm{B}\lambda\sigma} R^{(2)}_{\lambda\sigma} \rangle \qquad (4.62)$$

is (up to some factor) the averaged effective energy momentum tensor of the gravitational waves determined by (4.59). Now it is easy to see that the identification of the reference metric and the background metric leading to

$$R^{(0)}_{\mu\nu} = R^{\mathrm{R}}_{\mu\nu} \qquad (4.63)$$

excludes gravitational waves described by (4.59), because their energy—momentum tensor is equal to zero.* Therefore, $1/l^2$ must be much smaller than $1/L^2$ and the two metrics can never be identified.

In summary, we may state that Rosen's L_{gf} can be used for gauge fixing in quantum field theory only in the approximation where the back-reaction of gravitational perturbations on the gravitational background can be neglected. If we consider a higher-order approximation (back-reaction) or even the full Einstein—Hilbert action, Rosen's model cannot be used as a gauge breaking model. Accordingly, there are two ways out of this dilemma.

(i) One revives the standpoint that arbitrary generalizations of the Hilbert gauge condition leading to L_{gf} terms, which preserve coordinate covariance with respect to a reference metric $g^{\mathrm{R}}_{\mu\nu}$, are suitable gauging terms. Therefore, one has to prove for each L_{gf} separately if it allows an identification of $g^{\mathrm{B}}_{\mu\nu}$ and $g^{\mathrm{R}}_{\mu\nu}$. This must, of course, not only be done in the linear approximation but, at least, in an approximation regarding back-reaction. However, there are indications (see von Borzeszkowski, 1984; von Borzeszkowski and Treder, 1982b) that back-reaction leads to trouble for all background quantization schemes.

(ii) One takes the notion of gauge invariance and gauge breaking seriously and therefore the L_{gf} (4.28) can be looked at as a gauge-fixing term. Because in this case the Einstein—Hilbert action leads to the problems discussed above, one should start with a Lagrangian quadratic in the curvature. This can be considered as a further argument in favour of Adler's approach of induced action for GRT.

* We exclude contributions from ghost fields because of the interpretation of (4.62) as the energy—momentum tensor of gravitons.

Chapter 5

Quantum Postulates and the Strong Principle of Equivalence

Einstein's equations of GRT connect quantized non-gravitational matter described by its energy—momentum tensor $T_{\mu\nu}$ and gravitational fields described by the metric tensor $g_{\mu\nu}$ of a Riemannian spacetime. In order to avoid physical and mathematical inconsistencies resulting otherwise from Einstein's equations, one has to consider quantization of gravitational fields (see Chapter 1). The quantum procedure should unify or at least harmonize classical and quantum theory.* On the other hand, GRT is not genuine field theory. This is due to (i) the identification of gravitational field and spacetime metric (statement of the weak principle of equivalence) and (ii) the universal gravitational coupling making gravity itself a source of gravitational field (this is, together with (i), a formulation of the strong principle of equivalence). As a consequence of this strong principle, Einstein's equations show a typical nonlinearity producing back-reaction effects. It makes all quantization rules problematical which transform a usual classical field theory into a quantum field theory. In particular, one has to decide whether this quantization shows the existence of gravitons in the same sense as the physical existence of photons is considered to be verified. To discuss this problem, one must consider both quantized vacuum fields (together with the measurement of vacuum quantum effects) and the effects resulting from the coupling of gravitational fields to quantized non-gravitational matter.

5.1. Gravitons and the Linear Approximation of General Relativity Theory

First, let us look briefly at one of the usual quantum approaches (cf. von

* Harmonization requires that, according to the Bohr principle of correspondence, quantum field theory goes over into a classical theory for vanishing Planck's constant h. One finds here the same situation as for physical systems in the exterior gravitational field (cf. Chapter 1). Otherwise one would be led to Equation (1.3) implying the physically and mathematically senseless result that Newtonian gravitation depends on the Hilbert states.

Borzeszkowski, 1985). We start from the action integral

$$I = \frac{1}{\bar{\kappa}^2} \int R\sqrt{-g}\, \mathrm{d}^4x + \int \mathscr{L}_{\mathrm{M}}(\varphi)\, \mathrm{d}^4x, \tag{5.1}$$

where $\bar{\kappa}$ is the square root of the gravitational constant, $\mathscr{L}_{\mathrm{G}} = R\sqrt{-g}$ is the Lagrange density of Einstein and Hilbert, and $\mathscr{L}_{\mathrm{M}}(\varphi)$ is the Lagrange density describing the interaction of gravitation and matter fields φ. Using for conveniency the tensor densities

$$\tilde{g}^{\mu\nu} = \sqrt{-g}\, g^{\mu\nu},$$
$$\tilde{g}_{\mu\nu} = \frac{1}{\sqrt{-g}}\, g_{\mu\nu} \tag{5.2}$$

as basic fields, $\mathscr{L}_{\mathrm{G}}$ takes the simple Goldberg form

$$\mathscr{L}_{\mathrm{G}} = [2\tilde{g}^{\rho\sigma}\tilde{g}_{\lambda\mu}\tilde{g}_{\tau\nu} - \tilde{g}^{\rho\sigma}\tilde{g}_{\mu\tau}\tilde{g}_{\lambda\nu} - - 4\delta^{\sigma}_{\tau}\delta^{\rho}_{\lambda}\tilde{g}_{\mu\nu}]\tilde{g}^{\mu\tau}_{\ ,\rho}\tilde{g}^{\lambda\nu}_{\ ,\sigma}. \tag{5.3}$$

Just as in quantum electrodynamics, one can now make a perturbation theory by adding an appropriate gauge-breaking term and by assuming the ansatz

$$\tilde{g}^{\mu\nu} \to \delta_{\mu\nu} + \bar{\kappa}\phi_{\mu\nu} \tag{5.4}$$

which leads to an infinite series for $\tilde{g}_{\mu\nu}$:

$$\tilde{g}_{\mu\nu} = \delta_{\mu\nu} - \bar{\kappa}\phi_{\mu\nu} + \bar{\kappa}^2\phi_{\mu\alpha}\phi_{\alpha\nu} + \mathrm{O}(\bar{\kappa}^3). \tag{5.5}$$

Substituting (5.4) and (5.5) into (5.1), one obtains the series (written symbolically)

$$I = \int \text{“}\phi_{,\mu}\phi_{,\nu}\text{''} + \int \text{“}\varphi_{,\mu}\varphi_{,\nu}\text{''} +$$
$$+ \bar{\kappa} \int \text{“}\phi\phi_{,\mu}\phi_{,\nu}\text{''} + \bar{\kappa} \int \text{“}\phi\varphi_{,\mu}\varphi_{,\nu}\text{''} +$$
$$+ \bar{\kappa}^2 \int \text{“}\phi^2\phi_{,\mu}\phi_{,\nu}\text{''} + \bar{\kappa}^2 \int \text{“}\phi^2\varphi_{,\mu}\varphi_{,\nu}\text{''} + \cdots. \tag{5.6}$$

One can now proceed by discussing the different terms in (5.6). The parts proportional to $\bar{\kappa}^0$ determine the Lagrangian of free gravitational and matter fields. The term proportional to $\bar{\kappa}^1$ contains a part with one ϕ and two φ's, providing the diagram drawn in Fig. 5.1b, and a part like “$\phi\phi^2_{,\mu}$” implying a definite formula for the interaction of three gravitons, Fig. 5.1a. In the

Fig. 5.1. Feynman diagrams of the $\bar{\kappa}^1$-order approximation.

higher-order approximation, one obtains diagrams describing the gravitational Compton effect and so on; there arise also radiative corrections (closed loops) and hard renormalisation problems related to them.

Following this approach, one is nearly automatically led to the language of particle physics. We want however to rediscuss in this paper the question raised by Møller and Rosen, when Feynman (1963) asked at the Jablonna Conference on Relativistic Theories of Gravitation: is this theory really Einstein's theory of gravitation in the sense that, if you would have many gravitons, the equations would go over into the usual field equations of Einstein?

The answer to this question should not be trivial because the ansatz (5.4) implies more than the usual weak-field assumption. It confines all considerations to a region of dimension L_0, where one has a fixed flat background. In other words, one forces Einstein's GRT, as far as possible, into a special-relativistic field theory for a field $\phi_{\mu\nu}$. How strongly does GRT differ from usual special-relativistic theories if one forces GRT to behave similarly to special-relativistic theories? The field equations arising by this procedure are nonlinear, so, speaking in particle language, there arise graviton—graviton interactions against of a fixed background. This is, however, not the full nonlinearity of Einstein's equations which causes a back-reaction of the $\phi_{\mu\nu}$ field on the background and which realize the strong principle of equivalence.

To make the problem of linear approximation of GRT more evident, let us consider it in more detail (see von Borzeszkowski, 1982). We assume again that

$$g_{\mu\nu} = \gamma_{\mu\nu} + \varepsilon h_{\mu\nu}, \tag{5.7}$$

where $\gamma_{\mu\nu}$ represents the background field, say $\gamma_{\mu\nu} = O(1)$, $h_{\mu\nu}$ is a field of the same order of magnitude as $\gamma_{\mu\nu}$, and ε is a small parameter:

$$\gamma_{\mu\nu} = O(1), \; h_{\mu\nu} = O(1), \tag{5.8a}$$

$$\varepsilon \ll 1. \tag{5.8b}$$

Let us now introduce estimates of how fast the metric components vary by saying that typically their derivatives are of order

$$\partial\gamma \sim \gamma/L, \; \partial h \sim h/\lambda, \tag{5.9}$$

where L and λ are characteristic lengths over which the background γ and the h part of the field change significantly, without assuming any order-of-magnitude relation between L and λ from the very beginning. Then one may expand the Ricci tensor for the total metric in powers of ε to obtain (see Brill and Hartle, 1964; Isaacson, 1967; Misner *et al.*, 1973)

$$R_{\alpha\beta}(\gamma_{\mu\nu} + \varepsilon h_{\mu\nu}) = R_{\alpha\beta}(\gamma_{\mu\nu}) + \varepsilon R^{(1)}_{\alpha\beta}(h_{\mu\nu}) + \\ + \varepsilon^2 R^{(2)}_{\alpha\beta}(h_{\mu\nu}) + \varepsilon^3 R^{(3)}_{\alpha\beta}(h_{\mu\nu}) + \cdots, \tag{5.10}$$

where (the upright line denotes the covariant differentiation with respect to the background metric):

$$\begin{aligned} R_{\alpha\beta}(\gamma_{\mu\nu}) &= \text{the Ricci tensor of the background}, \\ R^{(1)}_{\alpha\beta}(h_{\mu\nu}) &= \gamma^{\rho\tau}(h_{\rho\tau|\alpha\beta} + h_{\alpha\beta|\rho\tau} - h_{\tau\alpha|\beta\rho} - h_{\tau\beta|\alpha\rho}), \\ R^{(2)}_{\alpha\beta}(h_{\mu\nu}) &= \tfrac{1}{2}[\tfrac{1}{2} h^{\rho\tau}_{|\beta} h_{\rho\tau|\alpha} + \\ &\quad + h^{\rho\tau}(h_{\rho\tau|\alpha\beta} + h_{\alpha\beta|\tau\rho} - h_{\tau\alpha|\beta\rho} - h_{\tau\beta|\alpha\rho}) + \\ &\quad + h^{\tau|\rho}_{\beta}(h_{\tau\alpha|\rho} - h_{\rho\alpha|\tau}) - \\ &\quad - (h^{\rho\tau}_{|\rho} - \tfrac{1}{2} h^{|\tau})(h_{\tau\alpha|\beta} + h_{\tau\beta|\alpha} - h_{\alpha\beta|\tau})] \end{aligned}$$

etc.

These terms have the following orders of magnitude:

$$\begin{aligned} R_{\alpha\beta}(\gamma_{\mu\nu}) &= O(L^{-2}), \\ \varepsilon R^{(1)}(h_{\mu\nu}) &= O(\varepsilon\lambda^{-2}), \\ \varepsilon^2 R^{(2)}_{\alpha\beta}(h_{\mu\nu}) &= O(\varepsilon^2\lambda^{-2}), \\ \varepsilon^3 R^{(3)}_{\alpha\beta}(h_{\mu\nu}) &= O(\varepsilon^3\lambda^{-2}) \end{aligned} \tag{5.12}$$

etc.

From this it becomes evident that only for

$$\lambda \gtrsim L \tag{5.13}$$

the powers in ε estimate completely the different terms, so that $R_{\alpha\beta}(\gamma_{\mu\nu})$ is dominant in magnitude. Accordingly, only for low-frequency fields $h_{\mu\nu}$, the background metric $\gamma_{\mu\nu}$ is governed by the equation

$$R_{\alpha\beta}(\gamma_{\mu\nu}) = 0. \tag{5.14}$$

The requirement

$$R_{\alpha\beta}(\gamma_{\mu\nu}) \stackrel{!}{=} 0 \text{ for all } \lambda \tag{5.15}$$

is thus a strong supplementary condition providing the new field equations

$$D_{\alpha\beta}(h_{\mu\nu}) \equiv \varepsilon R^{(1)}_{\alpha\beta}(h_{\mu\nu}) + \varepsilon^2 R^{(2)}_{\alpha\beta}(h_{\mu\nu}) + \\ + \varepsilon^3 R^{(3)}_{\alpha\beta}(h_{\mu\nu}) + \cdots = 0 \tag{5.16}$$

which differ, due to the nonlinearity of $R_{\alpha\beta}(\gamma_{\mu\nu} + \varepsilon h_{\mu\nu})$, from Einstein's field equations. Therefore, the supplementary condition (5.15) does, in general, not lead to a description of gravitational perturbations but defines the new field determined by the field equations (5.16). This $h_{\mu\nu}$ field runs through the curved $\gamma_{\mu\nu}$ background, without showing any back-reaction effect on $\gamma_{\mu\nu}$.

Accordingly, there exists in general only one possibility to interpret the results which one obtains via the ansatze given by (5.7), (5.8a, b) and (5.14): They must be considered as a weak-field and low-frequency approximation to the results which one expects for the exact field equations. An extrapolation to distances at which the field is strong and changes rapidly has to be viewed with caution. Of course, in the case of linear field equations such an extrapolation may also provide information about special types of strong and rapidly changing fields, because the linear equations corresponding to (5.14) and (5.16) have the same structure; in the linear case all depends on whether or not the expansion series in ε converges. But, in the case of nonlinear equations, an extrapolation of a solution obtained via the ansatze (5.7), (5.8a, b) and (5.14) to the whole λ scale is equivalent to the requirement (5.15) producing new, nongravitational equations.

It is now striking that the so-called background field method of quantizing gravity pioneered by DeWitt (1967) starts with the very ansatz (5.15). According to this method, the field $g_{\mu\nu}$ is decomposed as written in formula (5.7), where $\varepsilon h_{\mu\nu}$ is now a quantum field and $\gamma_{\mu\nu}$ the classical background field relative to which the quantum perturbations take place. Then the action $I(g_{\mu\nu}) = \bar{\kappa}^{-4} \int \sqrt{-g}\, R(g_{\mu\nu})\, \mathrm{d}^4x$ ($\bar{\kappa}^2$ is the Einsteinian gravitational constant) is expanded in a functional series about the background field. Later on, using the argument that the background field equations (5.15) are true when the external gravitons are physical, the validity of (5.15) is required as a basic constraint in this quantization procedure. (Under this assumption, all obtained formulas remain sensible of course if one sets $\gamma_{\mu\nu} = \eta_{\mu\nu}$; $\eta_{\mu\nu} \equiv$ Minkowski metric.)

The arguments given above at the level of the classical (nonquantized) field equations show however that such an approach reflects the properties of gravitation only for low frequencies, i.e., at comparatively large distances $\lambda \sim L$. Therefore, the results one obtains in this manner should not be extrapolated to small distances $\lambda \ll L$. Above all, the difficulties (as divergencies) which one obtains by extrapolating the usual background quantization conclusions to very small distances need not be physically disastrous.

It should be stressed here that, in the case of quantum gravity, the usage of the ansatz (5.7), (5.8a, b) and (5.14) is even more suspect than in purely classical considerations, at any rate if one has in view the quantum approach initiated by Feynman (1963) and Gupta (1968). Indeed, in the Feynman–Gupta approach, the operator $g_{\mu\nu}$ is separated into the classical Minkowski space background $\eta_{\mu\nu}$ plus a quantum correction $\varepsilon h_{\mu\nu}$, such that the Lorentz–Poincaré group allows us to perform the usual particle orientated quantum procedure. This procedure changes Einstein's general coordinate-covariant theory into a theory with new symmetries. It means therefore possibly a modification of the original physical content of Einstein's General Relativity Theory. Assuming (5.7) etc. within the framework of classical consideration, there arises of course an analogous problem. But then this method is used in order to calculate and to measure weak-field and low-frequency effects of a given theory. In quantum theory, the ansatz (5.7) etc. with $\gamma_{\mu\nu} = \eta_{\mu\nu}$ represents more than a method. By virtue of the fundamental meaning of the Lorentz—Poincaré group for quantum theory, it makes it only possible to formulate a quantum theory of gravity. Analogous arguments were true in the case of another fixed background if its symmetries were fundamentals of the quantization procedure. However, returning to the above given arguments, it is in general, at the classical as well as at the quantum levels, incorrect to extrapolate the results obtained by means of the ansatze (5.7), (5.8a, b) and (5.14) to arbitrarily small distances.

The above-described approach says therefore only that there are approximations for which one may use the concept of gravitons.* It remains to be determined whether this concept also has a physical sense for regions where the full nonlinearity of GRT has to be considered. To see what happens in this case, one should of course consider the full gravitational equations. Because of their complexity, we shall however start again with an approximate ansatz, namely the high-frequency approximation (Brill and Hartle, 1964; Wheeler, 1962; Isaacson, 1967; Choquet-Bruhat, 1969). This approximation considers the nonlinear effects of GRT in a more essential manner than the usual weak-field (low-frequency) methods. Accordingly, it should show at least some features of the nonlinearity of GRT.

Following this method and assuming accordingly that the effective energy density contained in a wave, $(c^4/G)(\varepsilon/\lambda)^2$, is of the same order of magnitude as $(c^4/G)L^{-2}$, the equation (5.14) must be replaced by $R_{\alpha\beta}(\gamma_{\mu\nu}) = \varepsilon^2 R^{(2)}_{\alpha\beta}(h_{\mu\nu})$.

Of course, it will be difficult to obtain rigorous results if one includes the back-reaction at the high-frequency quantum level. In subsequent sections some qualitative conclusions will be drawn about "high-frequency quantum gravity".

5.2. Gravitons and the Nonlinear High-Frequency Approximation of General Relativity Theory

From the high-frequency point of view one has to assume that the back-

* As long as one assumes a fixed background, this situation does not change essentially if one supposes, instead of a flat Minkowski background, some curved background.

ground curvature is equal to or greater than G/c^4 times the total energy density

$$\rho \approx \frac{c^4}{G} \frac{\varepsilon^2}{\lambda^2} \tag{5.17}$$

curving the background

$$L^{-2} \gtrsim \frac{G}{c^4} \frac{c^4}{G} \left(\frac{\varepsilon}{\lambda}\right)^2, \tag{5.18}$$

i.e.,

$$\varepsilon \lesssim \lambda / L. \tag{5.19}$$

(When no other sources of energy besides gravitational waves are present, the equality sign must be used.) According to condition (5.8), the following relation holds

$$\varepsilon \lesssim \lambda / L \ll 1, \tag{5.20}$$

so that this concept describes the propagation of a small-scale ripple in a background of a large-scale curvature. Here the case $\gamma_{\mu\nu} = \eta_{\mu\nu}$ is of course physically trivial because this assumption would lead to $L = \infty$ and, via (5.19), to the $\varepsilon = 0$ case. The back-reaction of the $\varepsilon h_{\mu\nu}$ field curving the $\gamma_{\mu\nu}$ background requires a vanishing $\varepsilon h_{\mu\nu}$ field for an uncurved background.

Let us now assume, in analogy to quantized electrodynamics, that the energy density contained in a gravitational wave of length λ is equal to*

$$\rho = \frac{\hbar \nu}{\lambda^3} = \frac{\hbar c}{\lambda^4}. \tag{5.21}$$

(Equation (5.21) results if one assumes that one graviton is contained in the volume $\sim \lambda^3$. For a generalization of this assumption, see below.) Together with equation (5.17) this quantization rule provides

$$\varepsilon = \frac{(\hbar G/c^3)^{1/2}}{\lambda}. \tag{5.22}$$

Consequently, looking from the viewpoint of the background quantization at the ansatz (5.7),

$$g_{\mu\nu} = \gamma_{\mu\nu} + \varepsilon h_{\mu\nu} \tag{5.23}$$

one must demand that $\varepsilon h_{\mu\nu}$ satisfies equation (5.22), while $\gamma_{\mu\nu}$ fulfils the

* Maybe, this assumption is not appropriate in order to quantize gravity. But without any assumption from quantum electrodynamics it is hardly possible to say what is meant by quantization. At any rate, it seems to be a minimal reference to quantum electrodynamics since it is used only to estimate the order of magnitude of ρ.

condition

$$\frac{(\hbar G/c^3)^{1/2}}{L} \ll |\gamma|. \tag{5.24}$$

The latter is the well-known correspondence condition which guarantees that the $\gamma_{\mu\nu}$ field may be considered as classical. For $\gamma_{\mu\nu} = O(1)$, it says

$$(\hbar G/c^3)^{1/2} \ll L, \tag{5.25}$$

i.e., that L has to be much greater than Planck's elementary length $l_P \equiv (\hbar G/c^3)^{1/2}$.

The comparison of (5.19) and (5.22) provides finally

$$\lambda \gtrsim (L l_P)^{1/2}. \tag{5.26}$$

From this inequality it is evident that the high-frequency and the quantum assumptions are only compatible for sufficiently large wavelengths λ.* In other words, our 'high-frequency background quantization' of gravity is only possible at distances much greater than Planck's elementary length l_P. For $L \to l_P$ the wavelength λ would become of the order of l_P, but this case is excluded by the classical background condition (5.25).

If one assumes now that the system to be quantized has the characteristic length $L_0 \gtrsim \lambda$, then one obtains (for an arbitrary particle number $n \geqslant 1$), instead of equation (5.21), the relation

$$\rho \approx \frac{\varepsilon^2 c^4}{G\lambda^2} \approx n \frac{\hbar\nu}{L_0^3} \tag{5.27}$$

leading to the quantization rule

$$\varepsilon = \sqrt{n}\,(\lambda/L_0^3)^{1/2}\, l_P \tag{5.28}$$

(for $n \gg 1$, equation (5.28) may be written as $\varepsilon \approx \tau^{1/2}\lambda^{1/2}$, where τ is the particle density). The comparison of the high-frequency condition (5.19) and of the quantization rule (5.28) provides then (see von Borzeszkowski, 1982, 1984; von Borzeszkowski and Treder 1982b)

$$\lambda \gtrsim \left[\left(\frac{\lambda}{L_0}\right)^{3/2} l_P L\right]^{1/2} \tag{5.29}$$

or

$$\lambda \gtrsim \frac{\bar{L}^2}{\bar{L}_0^3}\, l_P \tag{5.30}$$

* This corresponds to the relations of the electric field case, $\varphi = \varphi_0 + \varepsilon$ with $\varepsilon = (\hbar c)^{1/2}/\lambda$, $(\hbar c)^{1/2}/\lambda \ll \varphi_0$. However, in quantum electrodynamics there do not result such limitations for λ because it contains only the two universal constants $\hbar$ and c.

(with the dimensionless factors $\bar{L} = L/l_P$ and $\bar{L}_0 = L_0/l_P$). From (5.29) it becomes obvious that, in comparison to a system of the linear dimension $\sim \lambda$, there arises the factor $(\lambda/L_0)^{3/4}$. Accordingly, an increasing linear extension of the quantized system implies a decrease of the minimal λ value. The inequality relation (5.30) shows that, for a given background characterised by L, the quantized system must at least be of the linear dimension $L_0 \sim (l_P L^2)^{1/3}$ or $\bar{L}_0 \sim \bar{L}^{2/3}$, in order to allow a quantization of gravity also at distances of the order of magnitude $\sim l_P$.

The conditions (5.26) and (5.30) result from starting, on the one hand, with a classical background plus quantum fluctuation and requiring, on the other hand, that the classical background be influenced by back-reaction (via some average procedure). Now one could argue that the restricted compatibility of these two assumptions demonstrated by (5.26) and (5.30) shows that one should not simultaneously impose these requirements on gravity. Otherwise, there exist good physical arguments in favour of them. The first requirement rests on the fact that, for reason of measurement and physical interpretation, there should always exist a classical background; the latter is appropriate for including the back-reaction which it is necessary to consider for high-frequency fluctuations.

The same arguments are true for Einstein's equations modified at a distance l_P by terms resulting from the Lagrangian containing, beside the Einstein–Hilbert part R, the quadratic invariants R^2 and $R_{\mu\nu}R^{\mu\nu}$ (cf. von Borzeszkowski, 1982, 1984):

$$\begin{aligned} &R_{\alpha\beta} - \tfrac{1}{2} g_{\alpha\beta} R + l_P^2[\alpha \Box R_{\alpha\beta} + \\ &\quad + (\tfrac{1}{2}\alpha + 2\beta) g_{\alpha\beta} \Box R + \\ &\quad -(\alpha + 2\beta) R_{;\alpha\beta} + 2\alpha R_{\alpha\mu\nu\beta} R^{\mu\nu} + \\ &\quad -\tfrac{1}{2}\alpha g_{\alpha\beta} R_{\mu\nu} R^{\mu\nu} + 2\beta R R_{\alpha\beta} + \\ &\quad -\tfrac{1}{2}\beta g_{\alpha\beta} R^2] = \frac{G}{c^4} T_{\alpha\beta}. \end{aligned} \tag{5.31}$$

(α and β are numerical constants). Indeed, substituting (5.7) into the expressions for $R_{\alpha\beta;\rho\sigma}(\gamma_{\mu\nu} + \varepsilon h_{\mu\nu})$, one obtains the series

$$\begin{aligned} R_{\alpha\beta;\rho\sigma}(\gamma_{\mu\nu} + \varepsilon h_{\mu\nu}) = {} & R_{\alpha\beta|\rho\sigma}(\gamma_{\mu\nu}) + \\ & + \varepsilon R^{(1)}_{\alpha\beta\rho\sigma}(h_{\mu\nu}) + \varepsilon^2 R^{(2)}_{\alpha\beta\rho\sigma}(h_{\mu\nu}) + \cdots, \end{aligned} \tag{5.32}$$

where the right-hand terms have the following order of magnitude

$$\begin{aligned} &R_{\alpha\beta|\rho\sigma}(\gamma_{\mu\nu}) = O(L^{-4}), \\ &\varepsilon R^{(1)}_{\alpha\beta\rho\sigma}(h_{\mu\nu}) = O(\varepsilon\lambda^{-4}), \\ &\varepsilon^2 R^{(2)}_{\alpha\beta\rho\sigma}(h_{\mu\nu}) = O(\varepsilon^2\lambda^{-4}) \end{aligned} \tag{5.33}$$

etc. The terms

$$\Box R_{\alpha\beta},\ g_{\alpha\beta}\Box R,\ R_{;\,\alpha\beta} \tag{5.34}$$

provide expansion series in ε with terms of the same order of magnitude as in (5.32), while the quadratic terms $R_{\alpha\rho\sigma\beta}R^{\rho\sigma}$, $g_{\alpha\beta}R_{\rho\sigma}R^{\rho\sigma}$, $RR_{\alpha\beta}$ and $g_{\alpha\beta}R^2$ give expressions of the following type

$$R\ldots R^{\cdot\,\cdot\,\cdot} = \sim L^{-4} + \sim \varepsilon\lambda^{-2}L^{-2} + \sim \varepsilon^2 L^{-4} + \cdots \tag{5.35}$$

Considering now (5.10) and (5.12), one obtains on the left-hand side of (5.31) terms of the following orders

$$\begin{aligned} &\sim L^{-2} + \sim \varepsilon\lambda^{-2} + \cdots + \\ &+ l_P[\sim L^{-4} + \sim \varepsilon\lambda^{-4} + \sim \varepsilon\lambda^{-4} + \cdots]. \end{aligned} \tag{5.36}$$

By virtue of the conditions (5.26) this may finally be written as

$$\begin{aligned} &\sim L^{-2} + \sim l_P\lambda^{-3} + \cdots + \\ &+ l_P^2[\sim L^{-4} + \sim l_P\lambda^{-5} + \sim l_P^2\lambda^{-6} + \cdots]. \end{aligned} \tag{5.37}$$

Assuming again, according to the above-described procedure, that in the vacuum case this expression is proportional to the energy density of the waves times G/c^4, $\varepsilon^2\lambda^{-2} \sim l_P^2\lambda^{-4}$, we have

$$\begin{aligned} &L^{-2} \sim l_P^2\lambda^{-4}, \\ &l_P \sim \lambda^2 L^{-1}, \end{aligned} \tag{5.38}$$

i.e.,

$$\sim \frac{1}{\lambda L} + \sim \frac{1}{L^2} + \cdots + \left[\sim \frac{\lambda}{L^3} + \cdots + \sim \frac{\lambda^4}{L^6} + \cdots \right]. \tag{5.39}$$

The high-frequency condition $L \gg \lambda$ leads then to

$$\frac{1}{\lambda L} \gg \frac{1}{L^2} \gg \frac{\lambda}{L^3} \gg \frac{\lambda^2}{L^4} \gg \frac{\lambda^4}{L^6}. \tag{5.40}$$

Thus, the term of the nonmodified Einstein equations are dominant and we find essentially the same situation as before.

Considering now the measurement of gravitational fields against a curved background, one finds the same limitation (5.26) and (5.30) which followed from quantum field formalism (cf. von Borzeszkowski and Treder, 1982b). Indeed, if one wants to measure $h_{\mu\nu}$ field effects, then one has to consider perturbations $\delta\Gamma$ of $\Gamma \sim 1/L$ over a characteristic length λ

$$\delta\Gamma \sim \frac{\partial\Gamma}{L}\lambda \sim \frac{1}{L^2}\lambda. \tag{5.41}$$

The possibility of measuring such effects requires that $\delta\Gamma$ be greater than the uncertainty $\Delta\Gamma$ of the measurement,

$$\Delta\Gamma \leqslant \delta\Gamma \tag{5.42}$$

where $\Delta\Gamma$ satisfies the relation (3.46a) of Chapter 3. Therefore, one has

$$\delta\Gamma(L_0)^3 \sim \frac{\lambda}{L^2}(L_0)^3 \gtrsim \Delta\Gamma(L_0)^3 \gtrsim l_P^2 \tag{5.43}$$

and

$$\lambda \gtrsim \frac{1}{L_0^3}(l_P L)^2 \tag{5.44}$$

and, for optimal measurement,

$$\lambda \gtrsim (l_P L)^{1/2}. \tag{5.45}$$

Therefore, we find that the limitations (5.26) and (5.30) following from the quantum formalism are in accordance with the limitations resulting from Heisenberg's uncertainty relation for the measurement procedure (cf. relation (3.55)). In accordance with Pauli's argument (Pauli, 1933), such limitations on wavelength cannot occur in quantum electrodynamics, because wavelength depends on permeability. In gravitodynamics, this is excluded by the principle of equivalence, leaving no room for a changing gravitational permeability.

5.3. Compton Effect

As was shown above, there arise principle limitations in quantum gravity resulting from the problem of compatibility of the quantum principle and the fundamental principle of GRT, namely the principle of equivalence. This was shown by considering the quantum field formalism and the theory of measurement, respectively. In the quantum formalism, this problem of compatibility appears for high frequencies because of a contradiction between the quantum requirement (5.28) and the back-reaction implied by the equivalence principle and is brought into the calculation by the assumption $\varepsilon \lesssim \lambda/L$. The same was shown by the measurement discussion (see Chapter 3). There a contradiction shows up in the incompatibility of the Heisenberg uncertainty relation and the relation $M = m$ also being a consequence of the equivalence principle. According to Bohr and Rosenfeld, coincidence of the uncertainty relations derived from both formalism and measurement is necessary to prove the correctness of the quantum field formalism.

Do the relations (5.26) and (5.30), however, mean that such quantum effects as gravitational Compton effect, *Bremsstrahlung*, pair creation, and

Lamb shift should not exist? To answer this question (see von Borzeszkowski, 1985), let us return the expression (5.6) of the action functional I.

The term $\propto \bar{\kappa}^2$ in (5.6) (its gravitational part corresponds to the $R^{(3)}_{\alpha\beta}$ term in (5.10)) provides Feynman diagrams of the form shown in Fig. 5.2. And the term $\propto \bar{\kappa}^3$ in (5.6) (its gravitational part corresponds to the $R^{(4)}_{\alpha\beta}$ term in (5.19)) leads to diagrams of the type drawn in Fig. 5.3. One can see that the

Fig. 5.2. Feynman diagrams of the $\bar{\kappa}^2$-order approximation.

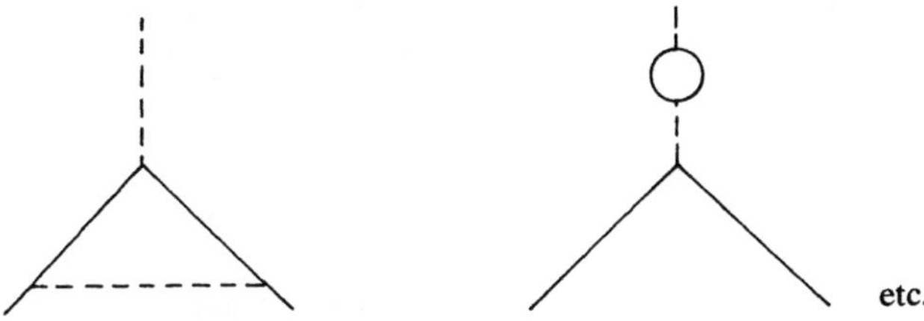

Fig. 5.3. Feynman diagrams of the $\bar{\kappa}^3$-order approximation.

cut-off length

$$\lambda \gtrsim (l_P L)^{1/2} \tag{5.46}$$

does not exclude all these effects. It only cuts off the high-frequency part of 'graviton—matter' interaction and pure 'graviton—graviton' interactions. Indeed, the Compton effect is, e.g., given by the formula

$$\frac{1}{\nu'} = \frac{1}{\nu} + \frac{\hbar}{mc^2}(1 - \cos\theta). \tag{5.47}$$

As was shown by Heisenberg (1938) in quantum electrodynamics, the introduction of a cut-off length r_0 restricts the validity of formula (5.47) to regions satisfying the relation

$$\lambda\Lambda \gg r_0^2, \tag{5.48}$$

where λ is the wavelength of the field and $\Lambda = \hbar/mc$ the Compton wavelength of the scattering particle. Assuming now, in accordance with (5.46), that $r_0 = (l_P L)^{1/2}$, one obtains for $\Lambda \approx L$:

$$\lambda \gg l_P. \tag{5.49}$$

This means the 'graviton—matter' Compton effect occurs for a matter-dominated background, where the selfinteraction of the gravitational field can be neglected. Assuming however 'graviton—graviton' interaction as the dominating process, i.e., assuming $\Lambda \approx \lambda$, then the relation (5.48) is not satisfied generally. Then $r_0 = (l_P L)^{1/2}$ cuts off such effects. The same can be shown for the higher-order diagrams.

In a theory with permeability μ, the Heisenberg condition (5.48) on the measurability of Compton effect is changed, because then $\Lambda \to \Lambda' = \mu\hbar/mc$. From this point of view, Einstein's principle of equivalence can be considered as a principle excluding gravitational permeability, such that limitation (5.48) cannot be undermined.

The fact that effects of quantized GRT are only cut off for high frequencies is due to the method used here. An exact treatment of the theory should show that, when strong nonlinearity comes into the calculations, it is not sensible to speak of gravitons and to look for corresponding effects (see Section 4.4).

Concluding this section, let us make here a remark on cosmological effects. For this purpose, let the length characterizing the background be of a cosmological order of magnitude

$$L \sim cT \sim \frac{\hbar c}{Gm^2}\,\frac{\hbar}{mc} \sim 10^{40}\,\frac{\hbar}{mc} \sim 10^{27}\ \text{cm}, \tag{5.50}$$

where $\omega = \hbar c/Gm^2 = N^{1/2} \approx 10^{40}$ is the square root of Eddington's number N of heavy particles in the Einstein cosmos. Then it follows that only for $L_0 \gtrsim 10^7$ cm may λ become equal to l_P, i.e., the system must have a dimension for which quantum effects are negligible. (For $L^* = cT^* \sim (\hbar c/Gm^2)^{1/2}(\hbar/mc) \approx 10^7$ cm, i.e., $T^* \approx 10^{-3}$ sec, one obtains $L_0 \gtrsim 10^{-6}$ cm.)

However, $\lambda \to l_P$ should not be required here to determine L_0, since we want to find whether λ may become l_P. Accordingly, one should determine L_0 by other physical arguments (e.g., by the linear dimension of the physical system which is considered from the quantum point of view) and ask what value the length λ may take. If we consider, e.g., strong interactions, L_0 is given by the nuclear radius $L_0 \approx 10^{-13}$ cm. Then only for $L \sim 10^{-3}$ cm (this value corresponds to the age of the world $T \lesssim 10^{-13}$ sec) may λ become l_P. This means that in the so-called hadronic phase of the hot-world model the quantization of gravity is not possible at such distances.

Therefore, the conditions (5.26) and (5.30) say that quantum effects of gravity are irrelevant for cosmological considerations. In the case of pure gravitons this becomes even more evident. Indeed, then (5.30) reduces to (5.26) because λ is the only characteristic length which can be put equal to L_0; then (5.25) states that λ has to be always much greater than l_P.

5.4. Lamb Shift

As was mentioned in the previous section, an investigation of the full general relativity theory should show that it does not make any difference if one considers the gravitational field as a quantized or a non-quantized field. To give further arguments corroborating this point of view, we shall discuss here the gravitational Lamb shift and, in the next section, the problem of the gravitational *Hohlraumstrahlung* (see Treder, 1979).

According to Einstein's principles of general relativity and equivalence, the general-relativistic value O of an observable is obtained from its special-relativistic value O_η by multiplication of O_η by corresponding concomitants of the metric tensor $g_{\mu\nu}$. If H_η is the special relativistic energy operator, then the general relativistic operator taking into consideration the influence of gravity reads

$$H = \sqrt{g_{00}}\, H_\eta . \tag{5.51}$$

Since the eigenvalues of energy are Einstein-shifted, the uncertainty Δg of the metric $g_{\mu\nu}$ leads to an uncertainty

$$\Delta H \approx \tfrac{1}{2}\Delta g H \tag{5.52}$$

of the energy value caused by gravity; according to (5.52), originally sharp spectrum lines are 'smeared over'. The line broadening is given by

$$\frac{\Delta H}{H} \approx \tfrac{1}{2}\Delta g \approx \frac{1}{c^2}\Delta\phi. \tag{5.53}$$

On the other hand, the contributions δH of the quantum effects of the gravitational field to the energy spectrum (following from quantum field theory) are, e.g., in the case of two particles of the same relativistic masses $m = E/c^2$, determined by natural powers $n \gtrsim 1$ of the 'Sommerfeld numbers'

$$\alpha = \frac{GE^2}{\hbar c^5} \approx \left(\frac{E}{E_0}\right)^2 \frac{Gm_0}{\hbar c}. \tag{5.54}$$

Therefore, the Lamb shift is given by

$$\delta O \approx \alpha^n H_\eta \approx \alpha^n H. \tag{5.55}$$

Of course, the greatest effects arise for ultrarelativistic particles because for these particles the rest masses m_0 are multiplied by great Lorentz factors E/E_0. Considering that the Compton wavelength reads $\lambda_0 = \hbar c/E_0$, we obtain the following value of the de Broglie wavelength,

$$L \approx \frac{\hbar c}{E} \approx \frac{E_0}{E}\frac{\hbar}{m_0 c} \approx \frac{E_0}{E}\lambda_0.$$

From this we obtain the gravitational Lamb shift:

$$\delta H \approx \left(\frac{E}{E_0}\right)^{2n} \frac{G\hbar}{c^3 L_0^2} H \approx \frac{G\hbar}{c^3 L^2} H. \tag{5.56}$$

The relative energy E of particles can be determined by measuring the de Broglie wavelength λ. Then, the error of length measurement $\Delta\lambda$ must not be greater than λ itself to define E reasonably, i.e., the position of the body of measurement must be fixed to an indeterminacy $\Delta\lambda = \Delta L_0$ satisfying the inequality

$$\Delta\lambda \approx L_0 \leqslant \lambda \approx \hbar c/E. \tag{5.57}$$

This leads to an estimation of the perturbation of the metric by the body of measurement:

$$\frac{G\hbar}{c^3 L^2} \leqslant \frac{G\hbar}{c^3(\Delta\lambda)^2} \leqslant \tfrac{1}{2}\Delta g \tag{5.58}$$

where Δg satisfies the relation

$$(\Delta g)^n \leqslant \Delta g \leqslant 1.$$

Therefore, (5.58) says that the uncertainty Δg caused by the fluctuations $\Delta\phi$ of the gravitational potential of the measurement body (i.e., by the fluctuations of the absolute value of the gravitational potential) is greater than the Lamb-shift-type quantum effects of the gravitational field:

$$\delta H \leqslant \Delta H \approx \tfrac{1}{2}\Delta g \cdot H. \tag{5.59}$$

If one wants to determine the masses m of the particles directly from the gravitational accelerations caused by the gravitational potential ϕ of a test body, then one must know its position with an accuracy greater than $\hbar c/E$:

$$\Delta L_0 \lesssim \lambda \approx \frac{\hbar c}{E}. \tag{5.60}$$

However, if the masses m of particles are not defined reasonably (in a measure-technical manner) then the assertion on the Lamb shift cannot be checked.

The term broadening $\Delta H/H$ caused by the undetermined perturbation potentials $\Delta\phi$ of the test bodies necessary to measure particle energies is always greater than the Lamb-shift-like term broadening $\delta H/H$ resulting from possible quantization of the gravitational field.

We find thus, in concordance with our results in Chapter 4, that there is always a prevailing (predominance) of test body quantum effects, and not only in the case of vacuum fluctuations and Lamb shift. But then there exists no difference empirically provable between the classical and quantum field theories of gravitation (at the very least as long as we do not consider free gravitons).

Thus the contents of the classical and the quantized theories of gravitation, as far as they can be checked physically, were not different with respect to their application to quantized matter fields. Therefore, one does not need more information about the gravitational field $g_{\mu\nu}$ than one obtains from the average-valued Einstein equations and the equations for the expectation values of the Einstein tensor, respectively.

The consistent application of Einstein's principles of equivalence and relativity together with Heisenberg's quantum-mechanical uncertainty relation leads to the result that it does not make any difference if one considers the metric $g_{\mu\nu}$ as a c-number or a q-number.

This agrees with the general expectations one has for 'nonlocal field theories' (according to Heisenberg). To prove our conjecture, it is necessary to research, in addition to gravitodynamical effects of the Lamb-shift type, also effects that are connected with gravitational radiation and the existence of free gravitons (in the sense of Dirac and others) (see Section 5.5).

5.5. Black-body Radiation

Let us consider now the problem of gravitational *Hohlraumstrahlung* in more detail (see Treder, 1979).

The algebraic properties of free gravitons corresponding to Dirac's quantization of the exact Einstein equations of gravitation are in principle the same as those which were formerly derived by Pauli and de Broglie for the linearized Einstein equations (see Tonnelat, 1965); their properties are evident from Einstein's theory (1918) of the linearized gravitational waves. The gravitational quanta, the 'gravitons' of Pauli and Dirac, are particles with vanishing rest mass and spin 2, which are completely transversely polarized, i.e., only the maximal spin values +2 and −2 are realized (Pauli, 1958) (see also Treder, 1963).

In quantum field theory, the spontaneous emission of radiation is governed by other laws as in the classical theory of radiation. For field quanta with vanishing rest mass there is an equilibrium of radiation in a closed cavity that, because of the great number of excited modes, is determined by Planck's radiation law both in quantum electrodynamics and in the linearized quantum gravitodynamics. According to the classical theory of radiation, the energy radiation is divergent due to its dependence on the number of oscillators $\sim \nu^2 \sim 1/\lambda^2$. Both for electrodynamics and for gravitational radiation the Rayleigh—Jeans law (with its ultraviolet catastrophe) would be valid in this case (see Planck, 1906).

Jeans (1911) proposed to understand the real energy-frequency distribution $u(\nu)$ of the black-body radiation described empirically by Planck's law as a quasi-stationary state in the sense of the classical theory of radiation (with Hertz' instead of Planck's oscillators), "in welchem die Erzeugung hochfrequenter Strahlung noch nicht bemerkt wurde, weil man es bei den gegebenen Versuchs-Bedingungen nicht mit einem wirklichen Gleichgewicht zu tun habe" (Bohr, 1966).

Therefore, according to J. Jeans, one could not empirically distinguish between classical and quantum black-body radiation because it did not form an equilibrium of radiation, but radiation maintains quasistationarily the insulated spectrum of energy $u_0(\nu)$.

The formation of an equilibrium occurs by means of a 'dust particle' as a catalyst (Planck, 1906). Planck, Poincaré, Einstein, and Bohr have shown that Jeans' argument (which was supported by Rayleigh) is not correct. The laws of the classical theory of radiation cannot be maintained in electrodynamics because, via this 'dust particle', there arises very quickly an equilibrium radiation in the cavity.

In gravitodynamics, the cross section σ for the radiation of gravitational radiation by particles is considerably smaller than the electrodynamical cross section $\sigma_0 \approx (e^2/m_e\sigma^2)^2 \approx l^2$ of light scattered by electrons with mass m_e and 'classical radius' l (of the order of Heisenberg's elementary length $l \approx \hbar c/E_0$, i.e., the Compton wavelength of baryons). The gravitational cross section of a particle of mass m is given by

$$\sigma \approx \left(\frac{GE}{c^4}\right)^2 \approx a^2 \approx \left(\frac{E}{E_0}\right)^2 \left(\frac{m_0}{c^2}\right)^2, \tag{5.61}$$

i.e., it is proportional to the square of its gravitational radius

$$a = \frac{GE}{c^4} = \frac{Gm}{c^2} = \frac{E}{E_0}\,\frac{Gm_0}{c^2}\,. \tag{5.61a}$$

Now, the distances r must always be greater than a to prevent the scattering particle from destroying the measurable spacetime structure by its own gravitational potential $\phi \sim uc^2/r$ (i.e., it must be guaranteed that $g_{00} > 0$).

Furthermore, the gravitational radius a must be smaller than the matter wavelength Λ of the scattering particle (Λ defines the radius of an ultra-relativistic particle):

$$a = \frac{GE}{c^4} \leqslant \Lambda \approx \frac{\hbar c}{E}, \quad a^2 \leqslant a\Lambda \leqslant \Lambda^2. \tag{5.62}$$

From this one obtains

$$a^2 \lesssim \frac{G\hbar}{c^3} \equiv l_P. \tag{5.63}$$

Thus, the cross section σ of the scattering of gravitational radiation by a particle is always given by

$$\sigma \approx a^2 \leqslant \frac{\hbar G}{c^3} \approx 10^{-36}\left(\frac{e^2}{E_0}\right)^2 \approx 10^{-40}\sigma_0, \tag{5.64}$$

where

$$\frac{\hbar c}{E_0} \approx 10^2 \frac{e^2}{E_0^2} \approx l \tag{5.64a}$$

is the Compton wavelength of a hadron and

$$\sigma_0 \approx \left(\frac{e^2}{E_0}\right)^2 \left(\frac{\hbar c}{e^2}\right)^2 \tag{5.64b}$$

is its electrodynamical cross section. Therefore, the gravitational cross section σ is at least 10^{40} times smaller than the electrodynamical cross section and the time elapsing until an equilibrium distribution of gravitational radiation will be formed by means of scattering by dust particles is at least 10^{40} times greater than in the electrodynamical case.

This means the 'gravitational radiation in a cavity' is, indeed, never in equilibrium, and the arguments given by Jeans are correct in this case. The gravitational radiation, being in a cavity, is in a quasistationary state; both according to the classical and the quantum theory of gravitational radiation, it remains in the spectral distribution $u_0(\nu)$, insulated in the beginning and not becoming 'black'. Therefore, there exists no measurable difference between classical and quantum behaviour. If the cavity has the microphysical dimension

$$V \approx l^3 \approx \left(\frac{\hbar c}{E}\right)^3 \approx 10^6 \left(\frac{e^2}{E_0}\right)^3 \tag{5.65a}$$

then the time interval

$$T \approx 10^{40} \hbar / E_0 \approx 10^{40} / c \tag{5.65b}$$

will elapse up to the formation of an equilibrium of radiation,* i.e., until the gravitational radiation is 'blackened' through its scattering by dust particles. This means at least a time of the order of the age of the world must elapse; therefore, the expansion of the universe also prevents the formation of the equilibrium of radiation in microscopic ranges.

This quasistationarity of the insulated spectrum of gravitational radiation corresponds to the nonlocalizability of the Einsteinian energy–momentum complex. The Kirchhoff and Planck laws of the thermal radiation use the conception of a local density of radiation and energy, which does not exist for gravitational radiation. The above-given estimations are a measure for this nonlocalizability of gravitational radiation.

The impossibility of distinguishing empirically, on the strength of the laws

* In electrodynamics this elapsing time would be $(l/c)\,(\hbar c/\rho^2)$.

of radiation, between quantized and classical gravitational radiation becomes plausible if one recalls the following (epistemological) arguments. The nonlocalizable Einsteinian energy—momentum t_μ^ν is identical with the canonical energy—momentum density

$$t_\mu^\nu = \tfrac{1}{2}\left\{\delta_\mu^\nu \mathscr{L} - \frac{\delta \mathscr{L}}{\delta[\sqrt{-g}\, g^{\rho\sigma}]_{;\,\nu}}[\sqrt{-g}\, g^{\rho\sigma}]_{;\,\mu}\right\} \tag{5.66}$$

which belongs to Einstein's Lagrangian

$$\mathscr{L} = \sqrt{-g}\, g^{\mu\nu}(\Gamma^\rho_{\sigma\mu}\Gamma^\sigma_{\rho\nu} - \Gamma^\rho_{\sigma\rho}\Gamma^\sigma_{\mu\nu}) \tag{5.67}$$

where $R_{\mu\nu} = 0$ describes the free gravitational field (Einstein, 1916; Einstein *et al.*, 1922).

The only tensors of second rank that could be identified with a (local) tensor of gravitational energy in the theory of general relativity are the Ricci tensor $R_{\mu\nu}$ and the Einstein tensor $G_{\mu\nu} = R_{\mu\nu} - \frac{1}{2} g_{\mu\nu} R$. Indeed, $G_{\mu\nu}$ could be interpreted as the 'metric energy—momentum tensor' of the gravitational field, as was proposed by Lorentz in 1916 and Levi-Civitá in 1917 (see Pauli, 1958). Then, one has (according to Hilbert and Lorentz)

$$\frac{\delta \mathscr{L}}{\delta g_{\mu\nu}} = \sqrt{-g}\, G_{\mu\nu} \tag{5.68}$$

this is the definition of the metric energy—momentum tensor $T_{\mu\nu}$. But, for free gravitational fields, $G_{\mu\nu}$ vanishes due to Einstein's gravitation equations

$$G_{\mu\nu} = R_{\mu\nu} - \tfrac{1}{2} g_{\mu\nu} R = 0$$
$$(\text{or } R_{\mu\nu} = 0) \tag{5.69}$$

such that, following Lorentz' proposal, the energy density $\sqrt{-g}\, G_{00}$ of a free gravitational field is identical with zero. For that reason, Einstein (1918) answered Lorentz and Levi-Civitá that, using the tensor (5.8) as metric energy—momentum tensor of the gravitational field, it would not be possible to ascribe energy to the free gravitational waves.

Quantum-mechanically speaking, according to Lorentz' definition of energy, there do not exist free gravitons.

If the question of quantization of free gravitational fields (i.e., of the existence of free gravitons) could be decided empirically and could be found to be true (this would be conceivable), then one would have the paradoxical situation that one could decide empirically between Einstein's 'canonical' definition (5.6) and Lorentz' 'metrical' definition (5.8) of the energy—momentum complex of free gravitational fields, namely, in favour of the canonical one. However, this is excluded by the impossibility in principle of distinguishing between the classical and quantum theories of general relativity and of detecting free gravitons.

According to our arguments, the question of quantization of the gravitational fields may be not a physically meaningful question. This statement may be in concordance with Einstein's point in his discussion with Bohr and Born and in his *Autobiographical Notes*. The general theory of relativity is a much more fundamental conception than the contrast between classical physics and quantum physics.

5.6. Historical Remarks: Black-body Radiation and Compton Effect

To finish this chapter, let us take a short look at the relation of black-body radiation to Compton's effect (see von Borzeszkowski, 1986).

Discussing today the graviton idea, many authors presuppose unreflectedly that there should exist gravitons simply because the particle concept was shown to be successful in physics. One should not forget however that already the photon, i.e., the analogue by which physicists are mainly guided when they are talking about gravitons, played a singular role in particle physics. Landau and Peierls attempted in 1931 (i.e., after the foundation of quantum electrodynamics) to show that the photon is a mathematical and not a physical concept. For many years, Planck and Bohr have rejected the photon hypothesis, and Einstein himself had always a critical attitude towards his own photon hypothesis. From 1905 to 1925, he was almost the only one who took the light-quantum hypothesis seriously, without however speaking of the photon as a particle until its existence was proved by the Bothe—Geiger experiment. And from 1925 until the end of his life he had a sceptical attitude towards quantum mechanics and quantum field theory as a fundamental unification of particles and fields. For a discussion of the history of the photon hypothesis, see Pais (1979).

Even if one accepts now the viewpoint that the existence of photons is theoretically predicted and experimentically discovered (not all authors share ths point of view), it is useful to rediscuss the history of the photon idea in order to show that all the arguments which speak for the corpuscular theory of the electromagnetic radiation are untenable for gravitational theory.

Kirchhoff's formula

$$E_\nu = \mathscr{f}(\nu, T) \tag{5.70}$$

for the radiation within "a space enclosed by bodies of equal temperature T, through which no radiation can penetrate" (*Hohlraumstrahlung*) represents the starting point of the electromagnetic (classical and quantum) radiation theory. All further developments leading to the Stefan—Boltzmann law, to Wien's and to Planck's radiation laws are nothing but steps to find Kirchhoff's function $\mathscr{f}$. Remembering this fact, one sees at once that great difficulties stand in the way of a gravitational radiation theory. Indeed, as was argued in Section 5.4, the fundamental principles of Einstein's General Relativity Theory prevent the definition of a '*Hohlraum*' (cavity) and of

Hohlraumstrahlung. Therefore, it should not be possible to establish a gravitational 'Wien law' or 'Planck law'.

Now one could make objections to the argument that this already shows that there should not be attributed gravitons to the gravitational field in the same sense as there are attributed photons to the electromagnetic field. One could say that there exists no necessary coupling between Planck's law of black-body radiation and Einstein's photon hypothesis, according to which parcels of energy and momentum can be attributed to free electromagnetic fields. Indeed, the first step in the direction of the photon hypothesis, namely Einstein's light-quantum hypothesis, according to which electromagnetic radiation sometimes behaves as if it consisted of parcels of energy, was formulated without using Planck's law. And, further, the existence of photons was finally verified by the proof that there occur individual Compton scattering events, i.e., without regard to a photon gas. Accordingly, it is useful to have a closer look at Einstein's derivation of his hypothesis and its gravitational analogue.

Einstein (1905) was led to his postulate by an analogy between electromagnetic radiation in the Wien regime, where the spectral energy density ρ of radiation of temperature T is given by Wien's ansatz

$$\rho(\nu, T) = \alpha\nu^3 \exp(-\beta\nu/T) \tag{5.71}$$

(Wien, 1896) and a classical ideal gas of n particles, where the entropy S obeys the volume-dependence law:

$$s(v, T) - s(v_0, T) = \frac{R}{N} \ln\left(\frac{v}{v_0}\right)^n \tag{5.72}$$

(R is the gas constant and N Avogadro's number). If one attempts now to use similar arguments to establish a gravity-quantum postulate, one fails here again at the beginning because, due to the strong principle of equivalence (universal coupling principle), one finds no operational definition of a black body. It is impossible to construct a cavity or another physical system which acts as a black body and allows us to define an equilibrium radiation density $\rho(\nu, T)$. Moreover, the light-quantum postulate results from a high-frequency consideration. Wien's law is the high-frequency approximation of Planck's law, since Planck's density function (Planck 1900a, b)

$$\rho(\nu, T) = \frac{8\pi h\nu^3}{c^3} \frac{1}{\exp(h\nu/kT) - 1} \tag{5.73}$$

contains Wien's law for

$$h\nu/kT \gg 1. \tag{5.74}$$

As was shown in Section 5.2, just this approximation is however not unlimitedly valid for quantized GRT. This fact results for the same reason that a gravitational *Hohlraumstrahlung* cannot be defined, namely because of the universal nature of gravity.

In discussing now not only the light-quantum and the gravity-quantum hypothesis, attributing energy parcels to the fields, but also the full photon and graviton postulates, one is led to the Compton effect as a crucial experiment of the particle idea.

The photon postulate ascribes to the electromagnetic field particles carrying energy $E = h\nu$ and momentum $p = h\nu/c$. From this result the following relations governing the kinematics for the scattering of a photon off an electron at rest (Compton, 1922; Debye, 1923)

$$\hbar\mathbf{k} = \mathbf{p} + \hbar\mathbf{k}', \tag{5.75}$$

$$\hbar c|\mathbf{k}| + mc^2 = \hbar c|\mathbf{k}'| + (c^2\mathbf{p}^2 + m^2c^4)^{1/2} \tag{5.76}$$

($\mathbf{k}$ and $\mathbf{k}'$ are the momenta of the photon before and after scattering respectively and, $\mathbf{p}$ is the momentum of the electron after scattering). These equations imply that the wavelength difference $\Delta\lambda$ between the final and the initial photon is given by

$$\Delta\lambda = \frac{\hbar}{mc}(1 - \cos\theta) \tag{5.77}$$

(θ is the photon scattering angle). Because this relation was found to be satisfied, the photon postulate was finally accepted by the physics community.

Here now arises the question: what does the gravitational Compton effect mean for the acceptance of the graviton concept? (Let us assume here that one is able to make such experiments.) To answer this question one must make some more remarks about the controversies surrounding the electromagnetic Compton effect.

After Compton's discovery, Bohr, Kramers and Slater (1924) made theoretical proposals concerning the interaction of radiation and matter which were to avoid the need to draw the photon conclusion from Compton's measurements. Bohr *et al.* wanted to protect the free electromagnetic field from quantization. All peculiarities of the radiation theory should not be due to the particle nature of free fields but to peculiarities of the interaction between the virtual field of radiation and the illuminated atoms. According to the Bohr—Kramers—Slater proposal, the energy of the field should change continuously and the energy of the atoms discontinuously. This proposal contradicts of course a general law of energy conservation. Consequently, the Bohr—Kramers—Slater answer was to abandon the conservation of energy and momentum for radiation transitions. The conservation law should not hold for individual elementary processes but only statistically, i.e., as an average over many processes. Accordingly, Compton's measurements on $\Delta\lambda$ should only refer to the average change of the wavelength, so that the conservation laws (5.75) and (5.76) are not tested for individual processes.

As is known, the Bohr—Kramers—Slater approach was refuted by the experiments of Bothe and Geiger (1924) and Compton and Simon (1925). Our discussion shows however that it should be difficult to use a gravitational Compton-effect measurement as an experimental verification of the existence of gravitons. In general relativity one has in general no exact laws of energy—momentum conservation; the situation described by the Bohr—Kramers—Slater proposal is here realized to some extent. A discussion of the gravitational Compton effect is thus only possible in an approximation, where relations of the form (5.75) and (5.76) are satisfied ($\mathbf{k}$, $\mathbf{k}'$ are then the energy—momenta of 'gravitons' and $\mathbf{p}$ that of a scattering mass). The Compton effect can accordingly not be used for a strong test of the graviton concept (see Section 5.3).

To summarize, both the gravitational analogues of electromagnetic *Hohlraumstrahlung* and Compton effect show that one cannot draw a conclusion about the existence of gravitons.

Chapter 6

Planckions

6.1. Heavy Gravitons

The electromagnetic field, as described by Maxwell and Faraday, and the gravitational field, as described by Newton and Einstein, are uniquely determined by some fundamental postulates showing simultaneously analogies and differences of both these macroscopic fields.

In the electromagnetic case these postulates are as follows:

(i) The field sources are charges and currents.

(ii) Photons possess no charge and, according to the $1/r$-dependence of static electric fields, no rest mass.

(iii) There exist no neutral currents.

(iv) There exist no magnetic monopoles.

The postulates (i), (iii) and (iv) are formulated in Ampère's and Faraday's laws. (ii) says that fields move with light velocity, and (i) and (iv) lead together with (ii) to a wave equation for the vector potential being Lorentz and gauge invariant.

In the gravitational case, more precisely, in GRT, one requires correspondingly:

(i′) The field sources are energy, momentum, pressure, and stress (not angular momentum and spin).

(ii′) Gravitons possess no energy—momentum.

(iii′) Gravity and inertia are locally equivalent.

The principle of equivalence (iii′) says that there is a universal influence of 'wattless' gravitational fields on all physical processes; it determines the coupling between gravity and matter in a unique manner. Together with (i′) and (ii′), it provides GRT. Requirement (i′) establishes the metrical (and not the canonical) energy—momentum tensor as a source term for gravity, and (ii′) and (iii′) specify the gravitational potential to be the metric $g_{\mu\nu} = g_{\nu\mu}$ of a Riemannian spacetime satisfying a general covariant wave equation (namely, Einstein's equations without cosmological term).

On the one hand, the postulates (ii′) and (ii) are equivalent. They demand in both cases that one should not attribute those properties to field quanta which characterize the matter sources of fields given by (i) and (i′). On the other hand, (ii′) corresponds to (iii) postulating the nonexistence of neutral currents.

Unified theory of electroweak interaction modifies Maxwell's electrodynamics so that (ii) as well as (iii) is replaced by the new postulate: There occur 'as many' massive as restmassless photons. Thus the current of massive photons is the source of the field with restmassless photons, and vice versa.* This physical feature of the electroweak interaction may be modelled by the Bopp—Podolsky fourth-order equations (see Section 4.1).

As was demonstrated in Chapter 4, quantum gravity theory with matter source terms leads, via the Heisenberg—Euler—Kockel approximation, automatically to GRT with fourth-order derivative corrections containing heavy gravitons, the latter figure as neutral currents. In this section we discuss some properties of those fourth-order equations (see Borzeszkowski *et al.*, 1978).

The Einstein equations with higher derivative corrections have recently received great attention, because several authors, especially Stelle (1977), have proved that these equations are one-loop renormalizable. Such equations revive early suggestions by Bach, Weyl, Einstein, and Eddington. More recently, Lanczos, Buchdahl, and Sexl *et al.* have treated those equations in order to unify gravitational and electromagnetic fields and/or to remove singularities from the classical vacuum solutions. The latter aspect has been strongly stressed (Treder, 1975a, 1977; Yourgrau *et al.*, 1979). It was shown there that the Hilbert—Einstein Lagrangian, R, supplemented by terms proportional to $R_{\mu\nu}R^{\mu\nu}$ and R^2, leads — in the linear approximations — to static spherical-symmetric solutions which are a sum of the Newtonian and the Yukawa potentials. They tend to a finite value at the origin $r = 0$ and represent the static analogues of the short-range as well the long-range parts of the gravitational potential $g_{\mu\nu}$.

Such equations may only be considered as gravitational field equations if they furnish, approximately at least, the Newton—Einstein vacuum for large distances. To this end, it is generally not sufficient to demand that the static spherical-symmetric solutions contain the Schwarzschild solution or the Newtonian potential in the linear approximation. It is necessary that the Schwarzschild solution corresponds to an interior solution for physically significant equations of state. In the case of the linearized field equations, this reduces to the requirement that there exist solutions with suitable boundary conditions which — for point-like particles — satisfy a generalized potential equation possessing a δ-like source. All disussions of higher-derivative field equations, the so-called gauge field equations of gravitation (Yang, 1974) as well as the fourth-order equations, must therefore be evaluated under the aspect of the coupling with non-gravitational matter and of their consequences at large distances.

* These massive photons are the W bosons.

The most general Lagrangian, containing the Hilbert—Einstein invariant as well as the quadratic scalars, reads as follows:

$$\mathscr{L} = \mathscr{L}_4 l^2 + \mathscr{L}_2 + \kappa\mathscr{L}$$
$$= \sqrt{-g}\,[(\alpha R_{\mu\nu}R^{\mu\nu} + \beta R^2)l^2 + R + \kappa\mathscr{L}_{\mathrm{M}}], \tag{6.1}$$

where α and β are numerical constants; l is a constant having the dimension of length. We use the signature $(+---)$ for the metric tensor $g_{\mu\nu}$. The Riemann tensor is defined by $R^{\lambda}_{\mu\alpha\nu} = \Gamma^{\lambda}_{\mu\alpha,\,\nu} + \cdots$ and the Ricci tensor by $R_{\mu\nu} = R^{\lambda}_{\mu\nu\lambda}$. Variation of the action integral $I = \int \mathscr{L}\, \mathrm{d}^4x$ results in the field equations of the fourth order, viz.,

$$l^2[\alpha\Box R_{\mu\nu} + (\tfrac{1}{2}\alpha + 2\beta)g_{\mu\nu}\Box R - (\alpha + 2\beta)R_{;\mu\nu} +$$
$$+ 2\alpha R_{\mu\alpha\beta\nu}R^{\alpha\beta} - \tfrac{1}{2}\alpha g_{\mu\nu}R_{\alpha\beta}R^{\alpha\beta} + 2\beta RR_{\mu\nu} +$$
$$- \tfrac{1}{2}\beta g_{\mu\nu}R^2] + (R_{\mu\nu} - \tfrac{1}{2}g_{\mu\nu}R) = \kappa T_{\mu\nu}. \tag{6.2}$$

These equations consist of 2 parts, namely, the fourth-order terms $\propto l^2$ stemming from the above quadratic scalars and the usual Einstein tensor, where

$$T_{\mu\nu} = \frac{1}{\sqrt{-g}}\,\frac{\delta\mathscr{L}_{\mathrm{M}}}{\delta g^{\mu\nu}} \quad (\text{with } T^{\nu}_{\mu;\,\nu} = 0). \tag{6.3}$$

A well known example of pure field equations of fourth order are the Bach—Weyl equations, derived from the Lagrangian (Weyl, 1919, 1923; Bach, 1921; Einstein, 1921; Lanczos, 1938)

$$\mathscr{L}^{\mathrm{B}}_4 = \sqrt{-g}\,(R_{\mu\nu}R^{\mu\nu} - \tfrac{1}{3}R^2), \tag{6.4}$$

proposed as a unified field theory of electromagnetism and gravitation with the property of conform-invariance in addition to Einstein's general covariance (see below).

In the framework of the Einstein—Bach—Weyl theory, with the Lagrangian

$$\mathscr{L} = \sqrt{-g}\,(R_{\mu\nu}R^{\mu\nu} - \tfrac{1}{3}R^2)l^2 + \sqrt{-g}\,R, \tag{6.5}$$

the conform-invariant Bach—Weyl tensor

$$B_{\mu\nu} = \frac{l^2}{\sqrt{-g}}\,\frac{\delta\mathscr{L}^{\mathrm{B}}_4}{\delta g^{\mu\nu}} \tag{6.6}$$

is proportional to the energy—momentum tensor of gravitons without the trace rest mass, because the trace B^{μ}_{μ} of $B_{\mu\nu}$ vanishes, while the Einstein tensor

$$R_{\mu\nu} - \tfrac{1}{2}g_{\mu\nu}R = \frac{1}{\sqrt{-g}}\,\frac{\delta\mathscr{L}_2}{\delta g^{\mu\nu}} \tag{6.7}$$

is the energy—momentum tensor of heavy gravitons. Thus, Einstein's vacuum equations $R_{\mu\nu} = 0$ state the vanishing of the energy—momentum of gravitons with rest mass: Einstein's theory does not allow the existence of heavy gravitons. On the other hand, the Bach—Weyl equations imply the vanishing of the energy—momentum tensor of the gravitons without rest mass. They must therefore not be considered as gravitational equations by reason of the fact that they do not describe long-range interactions.

In contrast, the Einstein—Bach equations

$$l^2 B_{\mu\nu} = -(R_{\mu\nu} - \tfrac{1}{2} g_{\mu\nu} R) \qquad (6.8)$$

assert that, roughly speaking, there always exist as many gravitons with rest mass as without. The energy—momentum tensor with rest mass, i.e., the Einstein tensor, is the 'source' of gravitons without rest mass; and, from $R_{\mu\nu} = 0$, one gets $B_{\mu\nu} = 0$, since $B_{\mu\nu}$ is a homogeneous function of $R_{\mu\nu}$. We have discussed the possibility of such an interpretation for the field equations (6.2) with arbitrary constants α and β.

To discuss these equations in more detail and to compare pure field equations of fourth order, e.g., the Bach—Weyl equations, with equations containing the Einstein tensor, we choose to consider the general static spherical-symmetric vacuum solutions of equations (6.8) in the linear approximation, especially the Green functions. Furthermore, we shall treat gravitational shock waves in order to make more manifest the dynamic content of such equations.

In the linear approximation for static fields with the Hilbert coordinate condition

$$\partial_\nu g^\nu_\mu = \tfrac{1}{2} \partial_\mu g^\tau_\tau$$

the pure and the mixed field equations of fourth order produce simple potential equations for all components of the metric tensor if $\alpha = -3\beta$ or $\alpha = -2\beta$. The pure and the mixed equations with point-like distribution of matter can then be written, respectively, as:

$$\Delta\Delta\varphi = -4\pi a \delta(\mathbf{r}), \qquad (6.9)$$

$$\Delta(\Delta - k^2)\varphi = -4\pi a k^2 \delta(\mathbf{r}). \qquad (6.10)$$

These equations replace the Poisson equation deduced from GRT. Δ is here the Laplace operator and $\Box = \partial^2/c^2\partial t^2 - \Delta$. The constant a denotes the mass of a point-like particle, the distribution of which is given by the Dirac delta-function δ; k is a reciprocal length $\propto 1/l$. Correspondingly, the Green functions read $\varphi = \tfrac{1}{2} ar$ and $\varphi = a/r - (a/r)e^{-kr}$.

On the other hand, the general spherical-symmetric solutions are given by

$$\varphi = \tfrac{1}{2} ar + b + \frac{c}{r} + \mathrm{d} r^2 \qquad (6.11)$$

and

$$\varphi = \frac{A}{r} + B + C\,\frac{e^{-kr}}{r} + D\,\frac{e^{kr}}{r}. \qquad (6.12)$$

a, b, c, d, and A, B, C, D are arbitrary constants of different dimensions. The solutions (6.11) and (6.12) differ from the Green functions, since they are solutions of operator equations which presuppose more general sources for point-particles than the monopole sources in equations (6.9) and (6.10). Indeed, insertion of the potentials (6.11) and (6.12) into the left-hand side of (6.9) and (6.10), yields

$$\Delta\Delta\varphi = -4\pi a\delta(\mathbf{r}) - 4\pi c\Delta\delta(\mathbf{r}) \tag{6.13}$$

and

$$\Delta(\Delta - k^2)\varphi = -4\pi(A + C)\Delta\delta(\mathbf{r}) - 4\pi k^2 C\delta(\mathbf{r}). \tag{6.14}$$

One immediately sees that there exists a greater manifold for field equations of fourth order than in the case of field equations of second order. In order to select the physically meaningful solutions from that manifold, we have to impose an additional condition upon the structure of sources: The point-particle is to be described by Dirac's delta function. That is, the distribution of mass must be expressed by a monopole density, as in the exterior Schwarzschild solution.

From (6.11) and (6.13) it is evident that — in the case of pure field equations of fourth order — this condition will reduce the manifold of solutions to the Green function $\propto r$. But this solution does not satisfy the correct boundary conditions at large distances, i.e., those field equations do not mirror the long-range Newtonian interactions. They are consequently at most field equations describing free fields, in other words, equations of a unified field theory in the sense of Weyl's (1919) or Eddington's (1953) ansatz.

However, if one starts from the mixed field equations, then (6.12) and (6.14) indicate that the aforementioned condition fixes the constants A, B, C, D such that the manifold of solutions (6.12) reduces to the Green function

$$\varphi = \frac{a}{r} - \frac{a}{r}\,\mathrm{e}^{-kr}$$

($A = -C = a$, $B = D = 0$). The potential φ contains the long-range Newtonian interaction. The coupling of the gravitational field with non-gravitational matter thus requires the mixed field equations in order to express gravitation adequately.

In the quantum-theoretical interpretation, virtual field quanta without rest mass belong to the Green function $1/r$; field quanta with the rest mass $m = k\hbar/c$ correspond to the Yukawa potential, that is, k^{-1} becomes a Compton wavelength. Hence, the Green function $(1/r)\,(1 - \mathrm{e}^{-kr})$ reflects a 'mixture of the same number of quanta with and without rest mass'. The short-range as well as the long-range terms $-(\mathrm{e}^{-kr})/r$ and $1/r$ have different signs. We thus attain renormalization of the potential φ at $r = 0$ (($\varphi(0)$ and $(\mathrm{d}\varphi/\mathrm{d}r)|_{r=0}$) are finite) and thereby achieve stabilization of classical particle models.

Quantum geometrodynamics leads to the assumption that the heavy graviton is a 'Planckion', i.e., that its mass is equal to Planck's fundamental mass $m_P = (\hbar c/G)^{1/2}$. This mass results from the condition that such a purely gravitational particle obeys $(Gm^2)/l_P = mc^2$, where l_P is again Planck's fundamental length $(\hbar G/c^3)^{1/2}$. For the 'Planckion', the gravitational radius (Schwarzschild radius) Gm/c^2, the Compton wavelength $\hbar/mc$ and the 'classical particle radius', are the same quantities.

We have hitherto dealt with the Green functions in the especially simple (generalized) potential equations (6.10). They occur for $\alpha = -3\beta = 1$ and $\alpha = -2\beta = 1$, the reason being that the linearized vacuum equations reduce to

$$l^2 \Box R_{\mu\nu} + R_{\mu\nu} = 0, R = 0$$

and

$$l^2 \Box G_{\mu\nu} + G_{\mu\nu} = 0.$$

Both cases are characterized in that they possess only one kind of particle with non-vanishing rest mass.

More generally, for arbitrary α/β, the Green function reads:

$$\varphi = \frac{a}{r} - \frac{4}{3}\frac{a}{r} e^{-k_2 r} + \frac{1}{3}\frac{a}{r} e^{-k_0 r}; \tag{6.15}$$

$k_2 = \alpha^{-1/2} l^{-1}$ and $k_0 = (-2(\alpha + 3\beta))^{-1/2} l^{-1}$ are the masses of two types of heavy gravitons. Here must be $\alpha \neq -3\beta$; for $\alpha = -3\beta$ one has $k_0 = k_2$.

The arguments regarding Green functions clarify a point discussed by Eddington (1953) and other authors: The general pure fourth-order equation $H_{\mu\nu} = 0$ contains the Einstein spaces $R_{\mu\nu} = \lambda g_{\mu\nu}$ (where λ is an arbitrary cosmological constant) as special solutions (Pauli, 1958). However, the Schwarzschild metrics are here not the exterior gravitational fields of spherical sources. The quasi-Newtonian potential $\propto 1/r$ in the general spherical-symmetric solution of the bipotential equation (6.9) is not the exterior potential of a spherical distribution of mass $\propto \delta(r)$.

The Einstein tensor $G_{\mu\nu}$ is characterized by the fact that it is the only tensor containing $g_{\mu\nu}$ and its first and second derivatives, and satisfying identically the equation

$$G^{\nu}_{\mu;\nu} = 0. \tag{6.16}$$

GRT thus selects the simplest expression of the field equations without requiring further conditions of symmetry. The physical argument enabling us to establish those field equations is the analogy with the Poisson equation of Newton's theory of gravitation. It defines the source of the gravitational field, namely, the active gravitational mass, by

$$M(=m) = \frac{1}{4\pi G}\oint \nabla_i \varphi \, dS^i.$$

The pure field equations of fourth order $H_{\mu\nu} = 0$ with arbitrary α and β are constructed according to the same scheme as Einstein's second-order equations. They are determined by the requirement that they are not to contain derivatives higher than fourth order and must fulfil the differential identity

$$\left(\frac{\delta \mathscr{L}_4}{\sqrt{-g}\ \delta g^{\mu\nu}} g^{\alpha\nu}\right)_{;\nu} = H^{\ \nu}_{\mu;\nu}$$

$$= [\alpha \Box R^{\ \nu}_{\mu} + (\tfrac{1}{2}\alpha + 2\beta)\delta^{\ \nu}_{\mu} \Box R -$$
$$- (\alpha + 2\beta) R^{\ \nu}_{;\mu} + 2\alpha R^{\nu}_{\mu\alpha\beta} R^{\alpha\beta} -$$
$$- \tfrac{1}{2}\alpha \delta^{\ \nu}_{\mu} R_{\alpha\beta} R^{\alpha\beta} + 2\beta R R^{\ \nu}_{\mu} - \tfrac{1}{2}\beta \delta^{\ \nu}_{\mu} R^{2}]_{;\nu} = 0. \qquad (6.17)$$

By reason of the differential identities (6.16) and (6.17) following from the general coordinate covariance, we get — as in the case of Einstein's equations — the dynamic equation

$$T^{\ \nu}_{\mu;\nu} = 0, \qquad (6.18)$$

for the pure field equations of fourth order

$$l^2 H_{\mu\nu} = \kappa T_{\mu\nu} \qquad (6.19)$$

as well as for the mixed field equations

$$l^2 H_{\mu\nu} + G_{\mu\nu} = \kappa T_{\mu\nu} \qquad (6.20)$$

Consequently, the principle of equivalence is also satisfied for those equations.

The requirement of general coordinate covariance furnishes in the case of fourth-order equations a variety of field equations depending upon the parameters α, β, whereas the Einstein equations are determined by this symmetry group up to the cosmological term $\lambda g_{\mu\nu}$. If one postulates that the field equations are invariant with respect to the conform transformation $g^*_{\mu\nu} = \Lambda(x^\lambda) g_{\mu\nu}$, one obtains

$$\alpha = 1, \beta = -\tfrac{1}{3}. \qquad (6.21)$$

The Bach—Weyl equations are moreover not only the sole equations of fourth order with conform invariance but also the sole conform-invariant general-relativistic equations that are at all conceivable.

GRT selects the simplest equations from the manifold of general covariant field equations, while the demand of conform invariance specifies the structure of those equations uniquely.

One can argue against invoking the conform-invariant equations that the trace of $H_{\mu\nu}$ ($\equiv B_{\mu\nu}$) vanishes identically. Accordingly, the trace $T = T^{\mu}_{\mu}$ of the matter tensor must also vanish such that the Bach—Weyl theory can merely describe matter without rest mass. This circumstance is closely connected with Einstein's argument against Weyl's theory, namely, that the actual physical measurement operations require calibrated rods and clocks because the metric cannot be related only to isotropic hypersurfaces.

That means that in the case $\alpha = 1$, $\beta = -\frac{1}{3}$, there are supportive reasons for the replacement of the pure by the mixed field equations. These supportive arguments pass beyond those which derived from the discussion of the spherical-symmetric solutions of the linear theory for arbitrary α and β.

Of course, the mixed Einstein—Bach—Weyl vacuum equations are not conform-invariant, since the invariance is violated by the term $G_{\mu\nu}$ being the source of the gravitons without rest mass. In other words, the 'supersymmetry', i.e., the additional invariance with respect to conform transformations, is broken down by $G_{\mu\nu}$ such that heavy besides light gravitons appear. The thus arising 'supergravitation' leads to the above-mentioned regularization of the selfinteraction of point-like particles.

One could perhaps expect such a regularization of quantized GRT, because it leads to a barrier of length measurements. The 'supergravitation' described by

$$l^2 B_{\mu\nu} + G_{\mu\nu} = \kappa T_{\mu\nu} \tag{6.22}$$

is tantamount to a phenomenological description of some aspects of the quantum structure of gravitation. Of course, such an interpretation is only possible if the fundamental length l is equivalent to Planck's length $(G\hbar/c^3)^{1/2}$. Only in this case l is not a new and *ad hoc* fundamental constant but derived from constants occurring in quantized gravitation.

The trace of the field equations (6.2) is as follows:

$$l^2(2\alpha + 6\beta)\Box R - R = \kappa T. \tag{6.23}$$

For the Einstein—Bach—Weyl vacuum equations ($\alpha = 1$, $\beta = -\frac{1}{3}$, $T = 0$), equation (6.23) yields

$$R = 0. \tag{6.24}$$

The vacuum field equations possess then the especially simple form

$$l^2(\Box R_{\mu\nu} + 2R_{\mu\alpha\beta\nu}R^{\alpha\beta} - \tfrac{1}{2} g_{\mu\nu}R_{\alpha\beta}R^{\alpha\beta}) + R_{\mu\nu} = 0 \tag{6.25}$$

which supply, for static spherical-symmetric fields, the above-discussed bipotential equations in the linear approximation. Those equations are characterized by one kind of particles with nonvanishing rest mass.

In order to discuss some aspects of the dynamic content of the fourth-order field equations in more detail, we now turn to the problem of shock waves. They enable us to arrive at conclusions regarding gravitons without quantizing the gravitational field. For a detailed discussion of massive shell models and shock waves, see Appendix A.

For the case of Einstein's vacuum field equations, the term containing linearly the second derivative of $g_{\mu\nu}$, viz.,

$$\tfrac{1}{2}\, g^{\alpha\beta}(g_{\mu\nu,\,\alpha\beta} + g_{\alpha\beta,\,\mu\nu} - g_{\alpha\mu,\,\beta\nu} - g_{\mu\beta,\,\alpha\nu}) \tag{6.26}$$

produces the result that the characteristic surfaces of those equations are isotropic hypersurfaces

$$z(x^{\lambda}) = 0$$

with

$$z_{,\mu} z_{,\nu} g^{\mu\nu} = 0$$

and that the bicharacteristic

$$z_{;\mu\alpha} z_{,\beta} g^{\alpha\beta} = 0$$

are null-geodesics. The perturbations of the gravitational field and the fronts of gravitational waves propagate accordingly with the local velocity of light. The field equations are only linear with respect to the second derivatives, and the term (6.26) contains the contravariant $g^{\mu\nu}$ (being a nonlinear function of the $g_{\mu\nu}$). Hence, the assertion that the gravitational waves possess the same velocity of propagation as light, holds only for infinitesimal perturbations and gravitational shock waves.

Contrary to Einstein's equations, the pure fourth-order field equations do not determine the shock waves uniquely. They contain such a great manifold of solutions that — interpreted as gravitational equations — they are, from a physical point of view, too vague. In consequence, one has to impose additional conditions upon the sources of gravitation (as was already demonstrated in the case of the static solutions of the linearized theory).

In order to determine shock waves describing discontinuity of second derivatives of $g_{\mu\nu}$, one has — in the case of any fourth-order field equations, $A_{\mu\nu} = 0$ — to postulate additionally the integral conditions

$$\int_D \int_D A_{\mu\nu}\, \mathrm{d}^4x\, \mathrm{d}^4x = 0$$

and

$$\int_D A_{\mu\nu}\, \mathrm{d}^4x = 0. \tag{6.27}$$

D is here an arbitrary four-dimensional domain of integration. Choosing a very flat cylinder as the domain of integration such that it contains any arbitrary finite part of the wavefront $z = 0$, one obtains from (6.27)

$$\int \int_{z=-\varepsilon}^{z=+\varepsilon} A_{\mu\nu}\, \mathrm{d}z\, \mathrm{d}z = 0; \tag{6.28}$$

ε is an arbitrary small quantity. For $\varepsilon \to 0$, (6.28) is a necessary and sufficient condition in the sense that the vacuum equations correspond to a theory containing only sources described in (6.13) and (6.14). It excludes in particular sources $\propto \nabla\delta$. The integral equation

$$\iiint_{z=-\varepsilon}^{z=+\varepsilon} A_{\mu\nu}\, dz\, dz\, dz = 0 \tag{6.29}$$

would, moreover, exclude sources $\alpha\nabla^2\delta$. Equation (6.29) is tantamount to a refinement of the conditions imposed upon (6.14) and to an equation specifying the discontinuities of the first derivatives of $g_{\mu\nu}$.

To deduce from (6.28) the conditions for the discontinuities of the second derivatives of $g_{\mu\nu}$, we invoke the method developed by Treder (1962). Let us assume that the hypersurface, where the second derivatives are discontinuous, is given by

$$z(x^\lambda) = 0$$

and that

$$p_\lambda \equiv z_{,\lambda}$$

is the normal vector of this hypersurface. The hypersurface $z = 0$ divides the four-dimensional spacetime V_4 into V_4^- $(z < 0)$ and V_4^+ $(z > 0)$. All derivatives lying in this surface are continuous such that only the derivatives of $g_{\mu\nu}$ pointing to V_4^- or V_4^+ are discontinuous. For sufficiently small z, the $g_{\mu\nu}$ in the neighbourhood of z are therefore given by the series

$$g_{\mu\nu}^+ = g_{\mu\nu}^- + \frac{1}{2!}\underset{2}{\gamma}_{\mu\nu} z^2 + \frac{1}{3!}\underset{3}{\gamma}_{\mu\nu} z^3 + \cdots \tag{6.30}$$

The discontinuities of the second derivatives of $g_{\mu\nu}$ are then equal to the difference of $g^+_{\mu\nu,\lambda\tau}$ and $g^-_{\mu\nu,\lambda\tau}$ constructed for $z \to 0$ and read

$$[g_{\mu\nu,\lambda\tau}] \equiv \lim_{z\to 0} (g^+_{\mu\nu,\lambda\tau} - g^-_{\mu\nu,\lambda\tau}) = \underset{2}{\gamma}_{\mu\nu} p_\lambda p_\tau . \tag{6.31}$$

Application of the series (6.30) results in the corresponding development of

$$(\Box R_{\mu\nu})^+, (\Box R)^+, \text{and } (R_{;\mu\nu})^+$$

for the domain V_4^+. The integral condition

$$\lim_{\varepsilon\to 0} \iint_{z=-\varepsilon}^{z=+\varepsilon} (l^2 H_{\mu\nu} + G_{\mu\nu})\, dz\, dz = 0 \tag{6.32}$$

gives thus for $\underset{2}{\gamma}_{\mu\nu}$ the relation

$$\alpha \underset{4}{A}_{\mu\nu} p^\rho p_\rho + (\tfrac{1}{2}\alpha + 2\beta) \underset{4}{A}{}^\alpha_\alpha p_\rho p^\rho g_{\mu\nu} - - (\alpha + 2\beta) \underset{4}{A}{}^\alpha_\alpha p_\mu p_\nu = 0, \tag{6.33}$$

where

$$\underset{4}{A}_{\mu\nu} = g^{\sigma\tau}(\underset{2}{\gamma}_{\mu\nu} p_\sigma p_\tau - \underset{2}{\gamma}_{\mu\tau} p_\sigma p_\nu - \underset{2}{\gamma}_{\nu\sigma} p_\tau p_\mu + + \underset{2}{\gamma}_{\sigma\tau} p_\mu p_\nu) = 2[R_{\mu\nu}]. \tag{6.34}$$

The condition (6.33) for the second-order discontinuities contains neither terms stemming from the nonlinear parts of $H_{\mu\nu}$, nor terms resulting from $G_{\mu\nu}$. These terms affect only the higher-order discontinuities. For $\alpha = -3\beta$, equation (6.33) lead, by reason of (6.24), to

$$p_\rho p^\rho \underset{4}{A}_{\mu\nu} = 0, \quad \underset{4}{A}{}^\mu_\mu = 0 \tag{6.35}$$

and for $\alpha = -2\beta$, to

$$p_\rho p^\rho (\underset{4}{A}_{\mu\nu} - \tfrac{1}{2} g_{\mu\nu} \underset{4}{A}{}^\alpha_\alpha) = 0. \tag{6.36}$$

Equations (6.35) and (6.36) differ — as in the case of Green functions — in that (6.35) represents an equation for the first approximation of the Ricci tensor and in that (6.36) is a relation for the first approximation of the Einstein tensor.

The trace of equation (6.33) is

$$(2\alpha + 6\beta) \underset{4}{A}{}^\mu_\mu p_\rho p^\rho = 0; \tag{6.37}$$

together with complete system of equation (6.33) it yields

$$p_\rho p^\rho = 0, \quad \underset{4}{A}{}^\mu_\mu = 0. \tag{6.38}$$

Equation (6.38) satisfy the relations (6.33) identically so that all discontinuities propagating with the velocity of light and fulfilling $\underset{4}{A}{}^\mu_\mu \equiv \underset{2}{\gamma}_{\sigma\tau} p^\sigma p^\tau = 0$, are compatible with the field equations of fourth order ($\alpha \neq -3\beta$).

Because Einstein's vacuum equations $R_{\mu\nu} = 0$ are always solutions of the fourth order equations (6.2) with $T_{\mu\nu} = 0$, one can add them to the integrated field equation (6.32) as follows:

$$\lim_{\varepsilon \to 0} \int \int_{z=-\varepsilon}^{z=+\varepsilon} [(l^2 H_{\mu\nu} + G_{\mu\nu})\, dz\, dz + G_{\mu\nu} = 0.$$

In this manner, $G_{\mu\nu}$ is introduced explicitly as a source of the light gravitons so that the second-order discontinuities satisfy the same relations as in GRT.

A different situation prevails in the case of the Einstein—Bach—Weyl equations. The influence of the Einstein tensor can already be noted in the second-order discontinuities. As a matter of fact, one obtains more solutions of equation (6.33) in the pure Bach—Weyl theory, since (6.37) is satisfied identically by $\alpha = -3\beta$. Equation (6.33) furnish accordingly two solutions:

1. $p_\rho p^\rho = 0$, $\underset{2}{\gamma}_{\mu\nu}$ is merely restricted by $\underset{4}{A}_{\mu\nu} = 0$.
2. $p_\rho p^\rho = 1$, $\underset{4}{A}{}^\mu_\mu \neq 0$, where $\underset{2}{\gamma}_{\mu\nu}$ is determined by

$$\underset{4}{A}_{\mu\nu} - \tfrac{1}{6}\,\underset{4}{A}{}^\rho_\rho g_{\mu\nu} - \tfrac{1}{3}\,\underset{4}{A}{}^\rho_\rho p_\mu p_\nu = 0. \tag{6.39}$$

Equation (6.39) possesses non-trivial solutions. The conform invariant Bach—Weyl equations lead therefore to a greater manifold of discontinuities than for the equations $\alpha \neq -3\beta$. Apart from the discontinuities propagating with the local velocity of light c, there exist shock waves with a velocity $v < c$.

The latter are only excluded by the Einstein tensor breaking down the conform invariance. Due to the Einstein tensor, an additional relation follows now from the trace equation $R = 0$ affecting the second-order discontinuities, namely, $2\underset{4}{A}{}^\rho_\rho = R = 0$. Inserting the latter into (6.39), we get $\underset{4}{A}_{\mu\nu} = 0$, i.e., an equation which has no genuine solutions: The addition of the Einstein tensor breaking down the conform invariance (supersymmetry) of the Bach—Weyl equations becomes therefore the source of the gravitons propagating with the velocity of light.

The fact that the Einstein—Bach—Weyl equations possess only one kind of particle with non-vanishing rest mass, reflects the physical fundamentals of the mixed fourth-order equations. We are now able to construct one mass from the elementary constants, $\hbar$, c, and G. Hence, the Einstein—Bach—Weyl equations ($\alpha = -3\beta$) enjoy a privileged position among the mixed fourth-order equations and may be conceived to be field equations of a theory of some kind of 'supergravitation'.

The investigation of shock-waves provides only the behaviour of dynamical perturbations either near graviton fronts moving with light velocity or near graviton fronts moving with velocity smaller than light velocity. To discuss some physical consequences of the mixture of massless and massive gravitons resulting from (6.2), in the remainder of this section we consider once more some aspects of high-frequency waves (see von Borzeszkowski, 1981). For this purpose, the method described in the previous chapter is used.

In GRT the effective energy density contained in the wave, $(c^4/G)\,(\varepsilon/\lambda)^2$, is, by virtue of Einstein's equations, related to the curvature of the background, L^{-2}, in the following manner:

$$L^{-2} \gtrsim \frac{G}{c^4}\frac{c^4}{G}\left(\frac{\varepsilon}{\lambda}\right)^2, \quad g_{\mu\nu} = \gamma_{\mu\nu} + \varepsilon h_{\mu\nu}. \tag{6.40}$$

(The other terms on the left-hand side describing the detailed structure of waves and starting with $\varepsilon\lambda^{-2}$ are suppressed here since, according to the high-frequency method, the energy of waves is to determine the background curvature, while these waves themselves result from the equations in the $\varepsilon\lambda^{-2}$ limit.) Then the most interesting case, that the background curvature is due entirely to the waves, i.e., $L^{-2} = (\varepsilon/\lambda)^2$, establishes a relation between ε and λ such that one is only concerned with one smallness parameter. If one regards L as a constant of order unity, then $O(\varepsilon) = O(\lambda)$.

Considering the fourth-order equations, the inequality (6.40) can be replaced by

$$\begin{aligned} &\varepsilon\lambda^{-2} + L^{-2} + \cdots + \\ &\quad + l^2(\varepsilon\lambda^{-4} + \varepsilon\lambda^{-3} + \varepsilon^2\lambda^{-4} + L^{-4} + \cdots) \gtrsim \varepsilon^2\lambda^{-2}. \end{aligned} \tag{6.41}$$

Before repeating now the order-of-magnitude arguments given above, we have to compare the Einstein term and the terms $\propto l^2$ on the left-hand side.* To this end let us consider the three cases: $l \gg \lambda$ ($\lambda = l^n$, $n \geqslant 2$), $l \ll \lambda$ ($l = \lambda^n$, $n \geqslant 2$), and $\lambda = l$. Then we see that the leading terms of the fourth-order part are of the order of magnitude $\varepsilon\lambda^{2/n-4}$, $\varepsilon\lambda^{2n-4}$, and $\varepsilon\lambda^{-2}$, while the Einstein tensor starts with $\varepsilon\lambda^{-2}$. Accordingly, the fourth-order terms are only for $\lambda = l$ of the same orders of magnitude as the Einstein tensor. If $l \ll \lambda$, i.e., if the high-frequency field has a wavelength considerably larger than the universal length l, then the l^2 modification of GRT has no physically significant effect. In turn, if the wavelength λ is smaller than $l(\lambda \ll l)$, then the fourth-order terms are the decisive ones.

In analogy to the discussion in GRT, a comparison of both sides of the fourth-order equations shows the following features:

(a) $l \gg \lambda$. From (6.41) it is obvious that, requiring $\varepsilon = \lambda^{1+1/n}$, ε can be specified such that $\varepsilon^2\lambda^{-2}$ is of the same order as l^2L^{-4} $(\sim \lambda^{2/n}L^{-4})$. However, then both these terms are of higher order in λ than L^{-2}. Therefore, high-frequency waves with $l \gg \lambda$ cannot determine the 'gross coarse' background; this background can only be established via the long-range Einstein term. If, for $l \gg \lambda$, distances smaller than l are considered, the l^2 part of the field equations is dominant and describes short-range fluctuations (possibly quantum fluctuations).

* We assume $L \sim O(1)$ and use, instead of l and λ, in the following the dimensionless parameters $\lambda' = \lambda/L$ and $L' = l/L$. For simplicity we omit the prime and write simply λ and l.

(b) $l \ll \lambda$. Here the results do not strongly differ from GRT. The waves with $l \ll \lambda$ mirror the large-distance behaviour of the fourth-order equations which is mainly determined by the Einstein tensor.

(c) $l = \lambda$. Then one obtains from (6.41) the relation

$$\varepsilon\lambda^{-2} + L^{-2} + \cdots + \\ + (\varepsilon\lambda^{-2} + \cdots + \lambda^2 L^{-2} + \cdots) = \varepsilon^2\lambda^{-2}. \tag{6.42}$$

In order to guarantee that the right-hand term $\sim \varepsilon^2\lambda^{-2}$ is of the same order of magnitude as L^{-2}, one has, as in GRT, to require $L\varepsilon = \lambda$. Just for 'ε-short' waves the full content of the fourth-order equations becomes relevant. Accordingly we shall consider this case in more detail.

We assume now that the metric may be represented by the ansatz

$$g_{\mu\nu}(x) = \gamma_{\mu\nu}(x) + \varepsilon h_{\mu\nu}(x, \varepsilon), \tag{6.43}$$

where

$$L\varepsilon = l = \lambda \ll 1, \tag{6.44a}$$

$$L = O(1),\ \gamma_{\mu\nu} = O(1),\ h_{\mu\nu} = O(1), \tag{6.44b}$$

$$\gamma_{\mu\nu,\alpha} = O(1),\ \gamma_{\mu\nu,\alpha\beta} = O(1), \tag{6.44c}$$

$$h_{\mu\nu,\alpha} = O(\varepsilon^{-1}),\ h_{\mu\nu,\alpha\beta} = O(\varepsilon^{-2}). \tag{6.44d}$$

The inverse metrics is then given by the expression

$$g^{\mu\nu} = \gamma^{\mu\nu} - \varepsilon\gamma^{\mu\rho}\gamma^{\nu\sigma}h_{\rho\sigma} + \varepsilon^2\gamma^{\mu\rho}\gamma^{\nu\lambda}\gamma^{\sigma\alpha}h_{\lambda\alpha}h_{\rho\sigma}. \tag{6.45}$$

Then one obtains from fourth-order gravitational equations, *in vacuo* to lowest order $O(\varepsilon^{-1})$,

$$l^2[\alpha\gamma^{\alpha\beta}R^{(1)}_{\mu\nu|\alpha\beta} + (\tfrac{1}{2}\alpha + 2\beta)\gamma_{\mu\nu}\gamma^{\alpha\beta}\gamma^{\sigma\tau}R^{(1)}_{\sigma\tau|\alpha\beta} - \\ - (\alpha + 2\beta)\gamma^{\sigma\tau}R^{(1)}_{\sigma\tau|\mu\nu}] + R^{(1)}_{\mu\nu} - \tfrac{1}{2}\gamma_{\mu\nu}R^{(1)} = 0, \tag{6.46}$$

and, defining

$$\varepsilon^2 K_{\mu\nu|\alpha\beta} \equiv \varepsilon^2(R^{(2)}_{\mu\nu|\alpha\beta} - S^{\rho}_{\mu\alpha|\beta}R^{(1)}_{\rho\nu} - S^{\rho}_{\nu\alpha|\beta}R^{(1)}_{\mu\rho} - \\ - S^{\rho}_{\mu\alpha}R^{(1)}_{\rho\nu|\beta} - S^{\rho}_{\nu\alpha}R^{(1)}_{\mu\rho|\beta} - S^{\tau}_{\mu\beta}R^{(1)}_{\tau\nu|\alpha} - S^{\tau}_{\nu\beta}R^{(1)}_{\mu\tau|\alpha} - \\ - S^{\tau}_{\alpha\beta}R^{(1)}_{\mu\nu|\tau})$$

and

$$S^{\mu}_{\beta\gamma} \equiv \tfrac{1}{2}g^{\mu\alpha}(\varepsilon h_{\alpha\beta|\gamma} - \varepsilon h_{\beta\gamma|\alpha} + \varepsilon h_{\gamma\alpha|\beta})$$

we have in the next order

$$\varepsilon^2 l^2[\alpha\gamma^{\alpha\beta}K_{\mu\nu|\alpha\beta} + (\tfrac{1}{2}\alpha + 2\beta)\gamma_{\mu\nu}\gamma^{\alpha\beta}\gamma^{\sigma\tau}K_{\sigma\tau|\alpha\beta} - \\ - (\alpha + 2\beta)\gamma^{\sigma\tau}K_{\sigma\tau|\alpha\beta}] + \varepsilon^2(R^{(2)}_{\mu\nu} - \tfrac{1}{2}\gamma_{\mu\nu}R^{(2)}) \\ = -(R^{(0)}_{\mu\nu} - \tfrac{1}{2}\gamma_{\mu\nu}R^{(0)}). \tag{6.47}$$

The equation (6.46) replace the equations $R^{(1)}_{\mu\nu} = 0$ of GRT. They are again equations for the propagation of the gravitational waves $h_{\mu\nu}$. Equation (6.47) show how the waves create the background curvature.

To analyse equation (6.46), we shall try to obtain, as for GRT, a solution to (6.46) of the form

$$\psi_{\mu\nu} \equiv h_{\mu\nu} - \tfrac{1}{2}\gamma_{\mu\nu}h = A_{\mu\nu}e^{i\phi}, \tag{6.48}$$

where $A_{\mu\nu}$ is a slowly changing real function of position and ϕ is a real function with a large first derivative but no larger derivatives beyond this to correspond to a slowly changing k_μ,

$$R^{(0)}_{\alpha\beta\gamma\delta} = O(1), A^{\mu\nu}_{\ ,\nu} = O(1),$$
$$k_\mu = \phi_{,\mu} = O(\varepsilon^{-1}), k_{\mu,\nu} = O(\varepsilon^{-1}). \tag{6.49}$$

(These assumptions are compatible with the conditions (6.44b)—(6.44d).)

As was shown by Isaacson (1967), one can impose on the equations $R^{(1)}_{\mu\nu} = 0$ the gauge conditions

$$\psi^{\mu\nu}_{\ |\nu} = 0, \psi = \gamma^{\alpha\beta}\psi_{\alpha\beta} = 0. \tag{6.50}$$

(The upright line denotes the covariant derivative with respect to $\gamma_{\mu\nu}$.) This can be done although these equations change their form under an arbitrary gauge transformation because this alters only higher-order corrections in ε. Imposing here the same gauge condition* and considering the relations

$$\gamma^{\mu\nu}A_{\mu\nu} = 0, ik_\beta A^{\alpha\beta} + A^{\alpha\beta}_{\ |\alpha} = 0 \tag{6.51}$$

which follow from (6.50), we obtain from (6.46),

$$[al^2 A_{\mu\nu}k_\rho k^\rho - A_{\mu\nu}]k_\rho k^\rho +$$
$$+ i[al^2(4A_{\mu\nu|\alpha}k^\alpha k_\rho k^\rho + 2A_{\mu\nu}(k_\rho k^\rho)_{|\alpha}k^\alpha] +$$
$$+ 2A_{\mu\nu|\beta}k^\beta + A_{\mu\nu}k_\rho^{\ |\rho} + \cdots = 0, \tag{6.52}$$

where the terms in the first and second brackets are of order $O(\varepsilon^{-2})$ and $O(\varepsilon^{-1})$.

To lowest order, it follows from (6.52) that

$$k_\rho k^\rho = 0 \tag{6.53a}$$

and, respectively,

$$k_\rho k^\rho = 1/l^2\alpha. \tag{6.53b}$$

In both cases, the second-order terms in (6.52) provide

$$2A_{\mu\nu|\beta}k^\beta + A_{\mu\nu}k_\beta^{\ |\beta} = 0. \tag{6.54}$$

We see that one obtains, in the case of the gauge (6.50), two particles,

* This confines us to the $\alpha = -3\beta$ case.

Einstein gravitons with vanishing rest mass and gravitons with non-vanishing rest mass $m \neq 0$. Both particles move, according to (6.54), as gravitons in GRT and photons in electrodynamics. Assuming $A_{\mu\nu} = A\mathrm{e}_{\mu\nu}$ (A is the amplitude, $\mathrm{e}_{\mu\nu}$ the polarization tensor) one has

$$(\ln A)_{,\beta} k^{\beta} + \tfrac{1}{2} k^{\beta}_{|\beta} = 0 \tag{6.55a}$$

and

$$\mathrm{e}_{\mu\nu|\beta} k^{\beta} = 0. \tag{6.55b}$$

The manner in which the relations (6.53) follow from the field equations makes evident that the gravitons with $m = 0$ (Einstein gravitons) do not result from the pure Einstein tensor appearing as an item in the fourth-order equations. They are due to the fact that the fourth-order terms have such a form that the vacuum solutions of Einstein's equations are solutions of the fourth-order equations, too.

To discuss (6.47) we assume, following the procedure of the high-frequency approximation, that the averaged approximate field determines the background curvature. Then (6.47) can be rewritten in the form

$$R^{(0)}_{\mu\nu} - \tfrac{1}{2}\gamma_{\mu\nu} R^{(0)} = -8\pi \langle (G/c^4) T_{\mu\nu} + L_{\mu\nu} \rangle, \tag{6.56}$$

where

$$\langle (G/c^4) T_{\mu\nu} \rangle = \frac{\varepsilon^2}{8\pi} \langle R^{(2)}_{\mu\nu} - \tfrac{1}{2}\gamma_{\mu\nu} R^{(2)} \rangle, \tag{6.57a}$$

$$\langle L_{\mu\nu} \rangle = \frac{\varepsilon^2 l^2}{8\pi} \langle \alpha \gamma^{\alpha\beta} K_{\mu\nu|\alpha\beta} + (\tfrac{1}{2}\alpha + 2\beta)\gamma_{\mu\nu}\gamma^{\alpha\beta}\gamma^{\sigma\tau} \times \times K_{\sigma\tau|\alpha\beta} + (\alpha + 2\beta)\gamma^{\sigma\tau} K_{\sigma\tau|\alpha\beta} \rangle. \tag{6.57b}$$

(The symbol $\langle \ldots \rangle$ denotes the average values of tensors.) Due to the relations

$$\begin{aligned} &\langle S_{\mu\nu\ldots|\rho\rho} \rangle = 0 \\ &\langle h_{\nu}^{\tau|\rho} h_{\rho\mu|\tau} \rangle = -\langle h_{\nu}^{\tau|\rho} h_{\rho\mu} \rangle + \mathrm{O}(\varepsilon), \\ &h_{\mu\nu|[\rho\tau]} \sim \mathrm{O}(\varepsilon^2) \end{aligned} \tag{6.58}$$

proved by Isaacson (1967), the term $\langle L_{\mu\nu} \rangle$ may be neglected; it serves as a source of higher-order corrections to the metric. (This is in accordance with our introductory remarks which have already shown that the $l^2 L^{-4}$ term does not contribute to the background curvature.) From $\langle L_{\mu\nu} \rangle \approx 0$ it follows that formally the background curvature is here determined as in GRT, namely by the averaged effective stress tensor $\langle R^{(2)}_{\mu\nu} - \tfrac{1}{2}\gamma_{\mu\nu} R^{(2)} \rangle$. Of course, the gravitational fields from which this tensor is constructed differ from GRT; they result from (6.46) and correspond to a mixture of different particles.

6.2. Planckions as Biggest Elementary Particles and as Smallest Test Bodies

Planckions were shown in the previous section as elementary particles arising in quantum gravity. More generally, one can characterize the situation as follows (see Treder, 1985, for Planck's units, cf. also Appendix B.): Max Planck introduced 'natural units' for mass (or energy), length (or time), charge, and temperature, which are defined by the universal constants of relativity (velocity of light c), quantum physics (Planck's constant $\hbar$), gravitation (Newton's gravitational constant G), and atomic theory (Boltzmann's constant k). Planck's quantities

$$m_P = \left(\frac{\hbar c}{G}\right)^{1/2}, \; l_P = ct_P = \left(\frac{\hbar G}{c^3}\right)^{1/2}, \; q_P = (\hbar c)^{1/2}, \tag{6.59}$$

and

$$\theta_P = \frac{1}{k}\left(\frac{\hbar c^5}{G}\right)^{1/2} \tag{6.60}$$

are interpreted in theoretical physics as the resulting limits for extreme ultrarelativistic energies in connection with the GUT (grand unification theory with special relativity) and super GUT (grand unification with general relativity). Planck's mass m_P corrresponds to the maximum energy E of a point-like elementary particle according to super GUT:

$$E = m_P c^2 = \left(\frac{\hbar c^5}{G}\right)^{1/2}, \tag{6.61}$$

and Planck's length l_P is the corresponding de Broglie wavelength

$$\Lambda \approx \hbar c/E = l_P. \tag{6.62}$$

We can also define l_P by the condition that it equals the gravitational radius Gm/c^2 of the same ultrarelativistic particle:

$$\Lambda \approx \frac{\hbar}{mc} = \frac{Gm}{c^2} \to m^2 = \frac{\hbar c}{G}. \tag{6.63}$$

The Planckian charge q_P is of the same order of magnitude as the charges of strong interactions. The hypothesis of GUT is that, in the domain of Planckian particle energies, all interactions become unified interactions with Planckian charges q_P as coupling constants. (Of course, these limits only give orders of magnitude.)

The suggestion that the Planckian mass m_P is the rest mass of a point-like particle implies the existence of 'Planckions'. This means the existence of elementary particles whose Compton wavelength $\hbar/m_0c$ equals their gravitational radius Gm_0/c^2 and with charge q equal to their gravitational mass $\sqrt{G}\,m_0$:

$$\frac{\hbar}{m_0c}=\frac{Gm_0}{c^2}\rightarrow m_0=\left(\frac{\hbar c}{G}\right)^{1/2}, \tag{6.64}$$

$$Gm_0^2=Gm_P^2=q^2\rightarrow q^2=\hbar c. \tag{6.65}$$

The Planckian temperature θ_P corresponds to an average particle energy of the Planckian order E:

$$k\theta_P\approx E=m_Pc^2. \tag{6.66}$$

Beyond this limit the conception of 'particles' becomes meaningless.

When discussing in Chapter 3 the measurement process in quantum field theory and especially in quantum gravity, Planckions were shown to be the smallest measurement bodies (see also Treder, 1985). To make this point more evident, let us add some further remarks.

In the special relativistic quantum theory we have the two constants $\hbar$ and c. Such theories (and accordingly also GUT) provide only the Planckian charge q_P and not 'natural units' for lengths and energies. The uncertainty relations (3.25) and (3.26) between field F and length L_0,

$$\Delta F\cdot L_0^2\gtrsim\frac{\hbar c}{q}\text{ and }\Delta F\cdot L_0^3\gtrsim\frac{\hbar}{c}\,\frac{q}{m} \tag{6.67}$$

say that there are no problems for the determination of a field because we are able to make q/m small and q large simultaneously. Here all uncertainties are consequences of the quantum-mechanical Heisenberg relation.

In GRT taken together with quantum theory (and accordingly in super GUT) we have however (3.46b),

$$\Delta gL_0^2\gtrsim l_P^2\approx\frac{\hbar G}{c^3}. \tag{6.68}$$

Assuming now that the relation

$$\Delta g\lesssim 1 \tag{6.69}$$

is satisfied for geometrical reasons, we obtain from (6.68)

$$L_0\gtrsim l_P, \tag{6.70}$$

i.e., l_P is the minimal length of a measurement body.

Multiplying (6.70) by c^2/G, one obtains furthermore

$$m\gtrsim m_P, \tag{6.71}$$

i.e., m_P is the smallest mass of a measurement body. We now turn to the quantity $q_P = (hc)^{1/2}$ (see von Borzeszkowski and Treder, 1983b). The fact that q_P is formed from the universal constants $\hbar$ and c signals that it has, similar to Planck's units l_P, m_P and θ_P, a fundamental physical meaning going beyond the borders of vacuum electrodynamics. This is, however, a non-trivial assertion. Indeed, in the case of charged mesons, for instance, further constants enter the theory which could decrease the value of the minimum charge of the measurement body. By a modification of the Bohr—Rosenfeld arguments, one sees however that $(\hbar c)^{1/2}$ is also in this case the minimum charge. This becomes obvious if one discusses the consequences of (3.24) for the case of a meson field.

Since the meson field F is characterized by the following relations where ν is the frequency, λ the wavelength, v the group velocity, u the phase velocity of the de Broglie's matter waves and M is the inertial mass of the corresponding mesons:

$$E = Mc^2 = \hbar\nu, \tag{6.72}$$

$$p = Mv = \hbar/\lambda \tag{6.73}$$

where

$$\lambda\nu = u = c^2/v, \ v \lesssim c, \ u \lesssim c, \tag{6.74}$$

and the field strength is given by

$$F \sim (E/V)^{1/2} \sim (\hbar u/\lambda^4)^{1/2}. \tag{6.75}$$

Assuming now $L_0 \sim \lambda$ and requiring

$$F \gtrsim \Delta F \tag{6.76}$$

one obtains, by virtue of (3.24) and (6.75),

$$\rho\lambda^3 \gtrsim \frac{\lambda}{\Delta x} \frac{\lambda}{uT} \left(\frac{u}{c}\right)^{1/2} (\hbar c)^{1/2}. \tag{6.77}$$

This is the Bohr—Rosenfeld formula modified by a factor $(u/c)^{1/2}$. The relation $uT \lesssim \lambda$ guarantees that $T \lesssim \tau$, where $\tau = \nu^{-1}$. By this one remains within the period of oscillation, and so the average procedure does not cancel all time changes of the field strength. If one repeats our arguments for the case of a measurement body of dimension $L_0 > \lambda$, this relation must be replaced by $uT \lesssim L_0$. Referring to the measurement body, the requirement $uT \lesssim L_0$ is in accordance with the fact that this body is a classical one showing no quantum or field relations between its ends.

Because of the relations

$$\lambda \gtrsim \Delta x, \lambda \gtrsim uT, u \gtrsim c \tag{6.78}$$

one has again

$$\rho L_0^3 \gtrsim \rho \lambda^3 \gtrsim (\hbar c)^{1/2}, \tag{6.79}$$

i.e., the charge ρL_0^3 of the measurement body is necessarily greater than Planck's charge $(\hbar c)^{1/2}$. In quantum gravity, this leads — as was mentioned already — to the requirement (4.5a).

Accordingly, Planck's units l_P, m_P and q_P characterize the minimal measurement body. As far as Planck's temperature θ_P is concerned, the Boltzmann constant k is connected with the gas constant $\mathscr{R}$ via

$$\mathscr{R} = \mathscr{L} k$$

where $\mathscr{L}$ is Loschmidt's number. Without presupposing the atomistic structure of matter, the number $\mathscr{L}$ is infinite and Boltzmann's constant becomes zero. Planck's temperature θ_P implies that each classical test body must have a temperature less than θ_P.

The 'Planckions' thus have two complementary meanings. Firstly, a Planckion may be an elementary particle with rest mass m. In this case the Planckions are the quanta of an ultrashort interaction (ultrastrong supergravity); the Planckions are big 'gravitons' which can only exist as free particles for Planckian energies E_P. Secondly, the Planckion is the smallest classical body which can serve as a standard for measurements. Ultra-Planckian elementary particles with masses $m > m_P$ are 'black holes' with

$$\frac{\hbar}{mc} < \frac{Gm}{c^2}. \tag{6.80}$$

However, for classical bodies with $m > m_P$ we have linear dimensions $x \geqslant Gm/c^2 \geqslant \hbar/mc$, and the gravitational radii Gm/c^2 have no physical meaning. This role of the 'Planckions' as the smallest test bodies is a Lorentz invariant property (see Appendix B).

These two meanings of Planckions result in the fact that one cannot distinguish between quantum and classical gravitational fields (see Chapters 4 and 5). We are not able to prove experimentally any quantum effect of gravitation by measurements with test bodies larger than Planckions because we then measure in regions in which theory does not provide significant quantum effects. On the other hand, Planckions are the smallest test bodies and the effects one can measure by means of them are of the same order of magnitude as the perturbations caused by the measurement body. The meaning of GRT is therefore — as Einstein pointed out — beyond the contrast between classical and quantum physics. In contrast to quantum electrodynamics, the region between $\mathscr{G}$ and $\mathscr{H}$ discussed in Section 3.2 vanishes in GRT.

As was argued by Markov (1980, 1981), this limit may be compared with the role of the vacuum light velocity in SRT. In special relativity one cannot overcome this c-border. Relativistic effects can be calculated from SRT mechanics corrections to classical mechanics signaling the existence of the border c. But one cannot arrive at a physically sensible theory by considering the transition of relativistic mechanics to velocities $v = c$.

In quantum gravity with matter coupling, one can also calculate quantum effects in certain approximate cases. One cannot however go over to the domain characterized by the dimensions of a Planckion. Moreover, the full theory of quantum gravity lies in this region characterized by the Planck units, which is seen for vacuum quantum gravity, because in this case the action of the nonlinearity is not suppressed by the matter action.

From this point of view, in GRT the particle language can only be used within the framework of certain approximations describing the interaction of gravitation and matter or, more generally, for some low-frequency approximations to GRT.

According to Heisenberg (1967), nonlocal field theories, i.e., quantum field theories with a 'smallest length' (elementary length l), should be equivalent to local field theories to which belongs quantum mechanically a space of state vectors (Dirac—Hilbert—Fock space) with an indefinite metric. However, negative expectation values do not have a physical meaning. Therefore, such a nonlocal field theory has, according to Heisenberg, a mathematical redundance with respect to its physical contents. In the theory of general relativity, Planck's elementary length now shows up as Heisenberg's restriction on length measurements. Thus, in this sense, the theory of general relativity is a nonlocal field theory where this nonlocality corresponds quantum mechanically to the existence of vacuum fluctuations $\Delta g_{\mu\nu}$ of the metric $g_{\mu\nu}$. Therefore, the quantum theory of gravity contains, as a necessity, less physics than the classical theory of general relativity, and this just means that there is no distinction between classical and quantum gravitational fields.

To summarise, the cut-off length arising in quantum gravity restricts its validity and, accordingly, the concept of gravitons to the range of weak-field and low-frequency approximation. There are some similarities between the situation in quantum gravity and in a theory realizing Heisenberg's programme formulated in his 1938 paper for quantum electrodynamics. Heisenberg introduced there a fundamental length r_0 *ad hoc*. This length was to cut off higher-order approximations of quantum electrodynamics to evade the problem of divergencies. There are, however, two differences with quantum gravity. First, such a length need not be introduced *ad hoc* but arises automatically in quantum gravity (this was already repeatedly noticed in the literature). Second, the limitations arising here automatically do not signal, as assumed for quantum electrodynamics by Heisenberg, the transition to a new type of interaction connected with particles of mass $\hbar/r_0 c$. In quantum gravity these particles are Planckions which show up a principle limit of the $(\hbar, c, G)$ theory.

From this point of view, effects such as the gravitational Casimir effect (see Birrell and Davies, 1982) and gravitational particle creation (see Zel'dovich and Starobinsky, 1971; Hawking, 1975) result from assuming very special boundary conditions and/or from the fact that one is calculating within the framework of quantum field theory in given curved spacetime, i.e., within an approximation neglecting back-reaction. Such effects can therefore only test the existence of 'linear gravitons'. They cannot answer any fundamental question of genuine quantum gravity.

6.3. Foam and Block Spaces

The limitations on measurement of gravitational fields are, due to the geometrical character of gravity, necessarily limitations on the size of spacetime regions over which the Bohr—Rosenfeld quantum field measurement procedure is, in principle, feasible. One way of expressing this fact is to say that supplementary to the well-known nonlocalizability of classical, i.e., non-quantized gravitational fields (making the $L_0 \to 0$ limit for lengths L_0 physically senseless), there exists a finite limit l_P on the localizability of quantized gravitational fields. This is owed to the fact that localizability is inconsistent with Heisenberg's uncertainty principle taken in conjunction with the basic tents of general relativity.

In accordance with this reading, there were early attempts to account for these measurement limitations in the fundamentals of gravitational theory, either by introducing a foam-like structure (Wheeler, 1962) or a pico-lattice structure (*Gitter-Welt*, according to Heisenberg) of spacetime (Lanczos, 1966, 1979). Within the context of such programmes, some authors speak nowadays of solving the localizability problem in a consistent manner. This is also done, e.g., in the framework of the 'stochastic quantum mechanics'. To make our point quite clear we want to make some comments.

One may of course attempt to anchor the Planck limitation on spacetime measurements in the basic structure of spacetime by introducing a lattice, foam or stochastic geometry. One must however not argue that it is necessary to regard the limitation resulting from measurement considerations in spacetime structure dynamics. Our short qualitative remarks on the high-frequency quantization of GRT have shown that quantum gravity provides the same limitations as the measurement discussion. Therefore, there is no need to change the whole quantum gravity theory so drastically as is done by the schemes just mentioned, except (i) if there are physical grounds to mistrust this occurrence of the l_P limit, which says that gravity is essentially classical, and/or (ii) if one may hope that a quantum field theory formulated on lattice, foam-like or stochastic spacetime evades the divergencies that plague the conventional approach of relativistic field theories. The first reason concerns gravitodynamics itself, the latter the theories of matter fields.

As for reason (i), one must state however once more that, as long as one considers a quantized gravitational theory containing $\hbar$, c and G as fundamental constants, the l_P limitation is an unavoidable result of the dynamical laws. It expresses a condition for the consistency of the different physical principles connected with $\hbar$, c, and G. If one wants to avoid this conclusion, one has only one possibility, namely to consider a theory where one of the constants is dropped. This may be done, e.g., by considering bimetric theories of gravitation violating the strong principle of equivalence (here G is 'less universal', because there is no universal action of gravity on the spacetime background) or by theories with 'induced constant G' (see, e.g., Adler, 1982). (It should not be possible to drop the constant $\hbar$, because one cannot find a method to harmonize quantum matter fields and gravitational theory without quantizing gravity.) However, as far as the consistency problem is concerned, a scheme which replaces G or $\hbar$ by a fundamental length (and considers G or $\hbar$ as derived) does not change the situation in principle. On the contrary, it establishes the appearance of l_P as a fundamental axiom of the theory.

The reason (ii) for introducing modified geometrical conceptions goes back to Klein, Hjelmslev and Menger. Their ideas on experience geometry (*Erfahrungsgeometrie*) or natural geometry (*natürliche Geometrie*) were later revived by March, Heisenberg, Ivanenko and others, also in order to solve the divergency problems of quantum electrodynamics. Any new attempt needs therefore only one thing, namely success in resolving the divergence problem.

Menger (1949) proposed to replace the Riemannian geometry by a geometry representing a combination of a geometry of pieces (*Geometrie der Stücke und Klumpen*) and a probability geometry (*Wahrscheinlichkeits-geometrie*). Regarding the finite extension of measurement bodies and the stochasticity of measurement, he proposed to attribute a distribution function to any two points of space which we call the distance of the points, and he asserted that we must work with this stochastic geometry to interpret the quantum uncertainties in relativistic regimes consistently. However there arises a question asked by Einstein (1949). What is the meaning of the stochastic structure of the spacetime?

To our mind, there are no arguments in favour of such structures. The fact that the measurement device (in particular, a test body) has to show some classical physical features and that measurement theory and quantum dynamics should give results which do not contradict one another (this was just demonstrated to be the justification for field quantization procedures) does not at all support Menger-type approaches. As a matter of fact, Bohr's and Rosenfeld's arguments did *not* pretend that the limitations on measurement arise from the classical apparatus. On the contrary, they could just show that the apparent new limitations on quantized electromagnetic field measurements result if one forgets about the classical features of the apparatus. In quantum electrodynamics the classical apparatus prevents limitations. The Bohr—Rosenfeld result for quantum gravitodynamics rather is that limitations result from the classical apparatus *plus* quantum interaction between apparatus and field system *plus* the strong equivalence principle.

Finally we mention an argument given by Heisenberg (1938). Let us consider, for this purpose, a particle of mass m, whose Compton wavelength is accordingly given by

$$\lambda = \hbar/mc. \tag{6.81}$$

Then one has in the ultra-relativistic approximation, where the velocity v of the particle is comparable to light velocity *in vacuo* c,

$$\lambda \sim \hbar/p \sim \hbar c/E \tag{6.82}$$

(p denotes momentum and E the energy of the particle). Heisenberg's uncertainty relation provides thus the relation

$$\Delta x \Delta p \sim \Delta x \Delta E/c \gtrsim \hbar \tag{6.83}$$

such that, in the ultra-relativistic approximation, Δx obeys the inequality

$$\Delta x \gtrsim \frac{\hbar c}{\Delta E}. \tag{6.84}$$

From this one sees — and this is the conclusion drawn by Heisenberg — that, for $\Delta x \sim \lambda$, the uncertainty ΔE is greater than E,

$$\Delta E \gtrsim E \tag{6.85}$$

so that the concept of rest energy of the particle, and thus the concept of the particle becomes dubious.

This is however just the situation one meets in the case of Planckions. Our uncertainty relation (6.70) says indeed nothing else but that the uncertainty of length measurements is of the order of the Compton wavelength of Planckions. The Planckion is therefore not a usual particle, but it confines (or cuts off) the spectrum of elementary particles.

This cut-off is the same as this one we found by extending quantum gravity to high frequencies. The arising limitations are due to the fact that the role of gravity as a spacetime background, on which the measurement devices are established, contradicts the unlimited interpretation of gravity as a usual physical field. The principle of equivalence brings about limitations on the physical meaning of quantum procedures applied to gravity. This guarantees the compatibility of gravity as a protophysical measurement-founded background and as a physical field. This compromise can only be reached up to a certain boundary; below this boundary quantized GRT does not exist and above this boundary quantum effects should be physically insignificant.

Appendix A

Massive Shell Models and Shock Waves in the Gravitational Theories with Higher Derivatives*

Gravitational equations containing higher than second derivatives of the metric have exercised the minds of a number of authors for different reasons. In this Appendix we shall be concerned with field equations which stem from the variational principle

$$I = \int [l^2(\alpha R_{\mu\nu} R^{\mu\nu} + \beta R^2) + R + \kappa L_{\text{M}}] \sqrt{-g}\, \text{d}^4x, \tag{A.1}$$

where α and β are numerical constants and l is a constant having the dimension of length. We use the signature $(+---)$ for the metric tensor $g_{\mu\nu}$. The Riemann tensor is defined by $R^{\lambda}_{\mu\nu\alpha} = \Gamma^{\lambda}_{\mu\nu,\alpha} + \cdots$, and the Ricci tensor by $R_{\mu\nu} = R^{\lambda}_{\mu\nu\lambda}$. Variation of the $g_{\mu\nu}$ results in the field equations

$$\begin{aligned} &l^2[\alpha \Box R_{\mu\nu} + (\tfrac{1}{2}\alpha + 2\beta) g_{\mu\nu} \Box R - (\alpha + 2\beta) R_{;\mu\nu} + \\ &\quad + 2\alpha R_{\mu\alpha\beta\nu} R^{\alpha\beta} - \tfrac{1}{2}\alpha g_{\mu\nu} R_{\alpha\beta} R^{\alpha\beta} + 2\beta R R_{\mu\nu} - \tfrac{1}{2}\beta g_{\mu\nu} R^2] + \\ &\quad + (R_{\mu\nu} - \tfrac{1}{2} g_{\mu\nu} R) = \kappa T_{\mu\nu}. \end{aligned} \tag{A.2}$$

In the early days of GRT, equations which derive only from the quadratic invariants $\propto R^2$ were considered in order to attempt a unification of gravitational and electromagnetic fields. In particular, Bach (1921) and Weyl (1919, 1923) have discussed the conformally invariant fourth-order equations ($\alpha = -3\beta$) and Eddington (1953) the $\alpha = -2\beta$ cases.

Later such equations were reconsidered from the viewpoint of a gravitational theory with phenomenological matter modifying the Einsteinian gravitation at small distances (Buchdahl, 1962; Pechlaner and Sexl, 1966). It was shown (Treder, 1975a) that in this case the Einstein–Hilbert part $\int R \sqrt{-g}\, \text{d}^4x$ must be necessarily included and that one has to impose

* For this, see Borzeszkowski and Frolov (1980).

supplementary conditions on the sources $T_{\mu\nu}$ being more restrictive than in GRT in order to restrict the manifold of solutions in a physically reasonable manner. The consequences for singularities of massive body and cosmological collapse were analysed in several papers.

The interest in the equations (A.1) was greatly increased by the fact that such higher-derivative terms naturally appear in the quantum effective action describing the influence of the vacuum polarisation of the gravitational field by gravitons and other particles (DeWitt, 1965, 1975; 't Hooft, 1974; Deser *et al.*, 1974; Cooper, 1974). The leading terms with the maximal number of derivatives at the one-loop level are of the following structure: $R_{\dots}(\ln \Box)R^{\dots}$, and the theory (A.1) may also be considered as a good approximation for this effective theory of quantum gravity. We assume in the following that l is equal to Planck's length l_{P}. The corresponding effective equations contain the highest derivatives in a linear way, and this renders it possible to use distribution theory to describe the possible jumps of metric derivatives associated with the matter source.

Using this method, free gravitational shock waves of the equations (A.2) were discussed (see Borzeszkowski *et al.*, 1978), and it was shown that quantum effects remove classical singularities resulting in GRT from gravitational collapse (Frolov and Vilkovisky, 1979). Generalizing those considerations, we perform here an analysis of the behaviour of massive shell models in the theory given by (A.1). This will show typical features of the coupling between matter and gravitation in theories with fourth-order derivatives.

The simplest way to take into consideration the source of a field is to use the thin massive shell model, i.e., to consider the case where matter of finite mass is distributed in a thin layer near some surface Σ. As the highest derivatives come always linearly into our equations, and as the coefficient functions before this term contain only second derivatives of the metric, the coefficient functions remain smooth when the thickness of the layer tends to zero. This means that the components of the curvature and energy–momentum tensors considered as a distribution, have well defined limits. The corresponding energy–momentum tensor may be written in the form

$$T_{\mu\nu} = \tau_{\mu\nu}\delta(\Sigma). \tag{A.3}$$

Here Σ is a three-dimensional surface formed by world lines of the points of the thin massive shell.

If the shell is non-null and p_μ is a normal vector satisfying the condition

$$p_\mu p^\mu = 1, \tag{A.4}$$

then $\delta(\Sigma)$ is a distribution defined by the equation

$$\delta(\Sigma) = p^\mu \partial_\mu \theta(x; \Sigma), \tag{A.5}$$

where

$$\theta(x;\Sigma) = \begin{cases} 1 \text{ for } x > \Sigma \\ 0 \text{ for } x < \Sigma \end{cases}.$$

The notation $x > \Sigma$ ($x < \Sigma$) is used for points x lying in the direction of p^μ (in the opposite direction) with respect to the Σ surface. The $\delta(\Sigma)$ function may be defined in an equivalent way by the relation

$$\int_V f(x)\,\delta(\Sigma)\sqrt{-g}\,\mathrm{d}^4x = \int_\Sigma f\sqrt{|\gamma|}\,\mathrm{d}^3x \tag{A.6}$$

which is to be satisfied for any test function $f(x)$. Here γ is the metric induced on Σ by g.

In the case when the shell is null, $\delta(\Sigma)$ can be defined as follows. Let $x^\mu(u)$ be the generators of Σ; u is an affine parameter along x^μ and the coordinates x^A ($A = 2, 3$) are taken to be constant on a given generator. The arbitrariness of this coordinate choice on Σ is

$$\begin{aligned} \tilde{x}^A &= \tilde{x}^A(x^A), \\ \tilde{u} &= \alpha(x^A)u + \beta(x^A). \end{aligned} \tag{A.7}$$

Let S_u be the two-dimensional section u = const. of Σ and Γ_u be a null surface generated by null rays orthogonal to S_u and lying out of Σ. The affine parameter along the null generators of Γ_u can be taken so that the conditions

$$z|_\Sigma = 0,\ p^\mu|_\Sigma = (g^{\mu\nu}\partial_\nu z)_\Sigma,\ p^\mu d_\mu = 1 \tag{A.8}$$

are satisfied. The coordinates (u, x^2, x^3) we choose out of Σ to be constant along null generators of Γ. In the so-defined coordinate system ('N coordinates') one has

$$g_{\mu\nu} = \begin{pmatrix} g_{00} & 1 & g_{0A} \\ 1 & 0 & 0 \\ g_{0A} & 0 & g_{AB} \end{pmatrix},\ g^{\mu\nu} = \begin{pmatrix} 0 & 1 & 0 \\ 1 & g^{11} & g^{1A} \\ 0 & g^{1A} & g^{AB} \end{pmatrix}. \tag{A.9}$$

We define the $\delta(\Sigma)$ distribution as follows

$$\delta(\Sigma) = \mathrm{d}^\mu\partial_\mu\theta(x;\Sigma).$$

In the coordinates (u, z, x^2, x^3) the equation for Σ is $z = 0$ and

$$\delta^N(\Sigma) = \delta(z).$$

The arbitrariness of the N coordinate choice (A.7) results in the following transformations

$$\tilde{p}^{\mu} = \alpha^{-1} p^{\mu}, \tilde{\delta}(\Sigma) \equiv \tilde{\delta}^{N}(\Sigma) = \alpha\delta(\Sigma),$$

so that $\tau_{\mu\nu} \rightarrow \tilde{\tau}_{\mu\nu} = \alpha^{-1}\tau_{\mu\nu}$ in the equation (A.3). This arbitrariness reflects the usual Doppler shift effects and can be excluded if one chooses a special observer system t^{μ} to put $t^{\mu}p_{\mu} = 1$.

In order to analyse the structure of discontinuities of the derivatives of $g_{\mu\nu}$ induced by a thin massive shell which are compatible with the equations (A.1), we invoke the method developed by Papapetrou (1962) and Treder (1962). We assume $\alpha \neq 0$ and $\beta \neq 0$ and discuss jumps of the order with $n > 0$. The $\alpha = 0$ case may be obtained in general by specialising the $\alpha + 3\beta \neq 0$ case considered below.

Accordingly, we assume that the hypersurface Σ: $z(x^{\lambda}) = 0$, where the derivatives of the metric are discontinuous, divides the four-dimensional space-time V_4 into V_4^- ($z < 0$) and V_4^+ ($z > 0$). All derivatives lying in this surface are continuous such that only derivatives pointing to V_4^- or V_4^+ are discontinuous. For sufficiently small z, the $g_{\mu\nu}$ in the neighbourhood of $z = 0$ are therefore given by the series

$$g_{\mu\nu} = g_{\mu\nu}^- + \underset{n}{\gamma}_{\mu\nu}\, {}_+z^n + \underset{n+1}{\gamma}_{\mu\nu}\, {}_+z^{n+1} + \cdots. \tag{A.10}$$

Here $g_{\mu\nu}^-$ is the metric in V_4^- and its continuation in V_4^+ is four times differentiable. The $\underset{i}{\gamma}_{\mu\nu}$ are functions which are n times differentiable, and the distribution ${}_+z^n$ is defined by

$$\ldots, {}_+z^{-1} \equiv \delta(z), {}_+z^0 \equiv \theta(z), {}_+z \equiv z\theta(z), \ldots,$$

$${}_+z^n \equiv \frac{1}{n!} z^n\theta(z) \text{ for } n \geqslant 0. \tag{A.11}$$

The jumps of the derivatives of $g_{\mu\nu}$ are thus reduced to discontinuities of the ${}_+z^n$.

From (A.10) and (A.11) it follows that the jumps of the derivatives of $g_{\mu\nu}$ may be expressed by the functions $\underset{n}{\gamma}_{\mu\nu}$. For instance, if jumps of the first derivatives of $g_{\mu\nu}$ occur, they read

$$[g_{\mu\nu,\lambda}] \equiv \lim_{z\to 0} (g^+_{\mu\nu,\lambda} - g^-_{\mu\nu,\lambda}) = \left[\frac{\partial g_{\mu\nu}}{\partial z}\right] p_\lambda = \underset{1}{\gamma}_{\mu\nu} p_\lambda. \tag{A.12}$$

Using the differentiation rules

$$({}_+z^n)_{;\alpha} = ({}_+z^n)_{,\alpha} = p_\alpha\, {}_+z^{n-1},$$

we obtain from (A.10) for $R_{\mu\nu}$, R, $\Box R_{\mu\nu}$, $\Box R$, and $R_{;\mu\nu}$ the corresponding series:

$$R_{\mu\nu} = R^-_{\mu\nu} + \underset{n-2}{A}{}_{\mu\nu}\,{}_+z^{n-2} + \underset{n-1}{A}{}_{\mu\nu} z^{n-1} + \ldots,$$

$$R = R^- + \underset{n-2}{A}\,{}_+z^{n-2} + \underset{n-1}{A}\,{}_+z^{n-1} + \ldots,$$

$$\Box R_{\mu\nu} = (\Box R_{\mu\nu})^- + \underset{n-2}{A}{}_{\mu\nu} p^2{}_+z^{n-4} +$$
$$+ [\underset{n-2}{A}{}_{\mu\nu} p^\alpha_{;\alpha} + 2\underset{n-2}{A}{}_{\mu\nu;\alpha} p^\alpha + \underset{n-1}{A}{}_{\mu\nu} p^2]{}_+z^{n-3} + \ldots,$$

$$\Box R = (\Box R) + \underset{n-2}{A}\, p^2{}_+z^{n-4} +$$
$$+ [\underset{n-2}{A}\, p^\alpha_{;\alpha} + 2\underset{n-2}{A}{}_{,\alpha} p^\alpha + \underset{n-1}{A}\, p^2]{}_+z^{n-3} + \ldots,$$

$$R_{;\mu\nu} = (R_{;\mu\nu})^- + \underset{n-2}{A}\, p_\mu p_\nu z^{n-4} + \cdots +$$
$$+ [\underset{n-2}{A}\, p_{\mu;\nu} + \underset{n-2}{A}{}_{,\mu} p_\nu +$$
$$\underset{n-2}{A}{}_{,\nu} p_\mu + \underset{n-1}{A}\, p_\mu p_\nu]{}_+z^{n-3} + \ldots, \tag{A.13}$$

where

$$p^2 \equiv p^\rho p_\rho,\ \underset{n-2}{A} \equiv \underset{n-2}{A}{}^\rho_\rho,\ \underset{n}{\gamma} \equiv \underset{n}{\gamma}{}^\rho_\rho$$

and

$$\underset{n-2}{A}{}_{\mu\nu} = \tfrac{1}{2}(-p^2 \underset{n}{\gamma}{}_{\mu\nu} + \underset{n}{\gamma}{}_{\mu\rho} p^\rho p_\nu + \underset{n}{\gamma}{}_{\nu\rho} p^\rho p_\nu - \underset{n}{\gamma}\, p_\mu p_\nu).$$

From now on the operation $B_\mu \rightarrow B^\mu$ and the covariant differentiation are formed by the $(g_{\mu\nu})^-$ and the $(\Gamma^\lambda_{\mu\nu})^-$ in the negative domain V_4^-. Inserting (A.13) into (A.1), we obtain

$$l_P^2\{[\alpha \underset{n-2}{A}{}_{\mu\nu} p^2 + (\tfrac{1}{2}\alpha + 2\beta) g_{\mu\nu} \underset{n-2}{A}\, p^2 - (\alpha + 2\beta) \underset{n-2}{A}\, p_\mu p_\nu]{}_+z^{n-4} +$$
$$+ [\alpha(\underset{n-2}{A}{}_{\mu\nu} p^\alpha_{;\alpha} + 2\underset{n-2}{A}{}_{\mu\nu;\alpha} p^\alpha + \underset{n-1}{A}{}_{\mu\nu} p^2) +$$
$$+ (\tfrac{1}{2}\alpha + 2\beta) g_{\mu\nu}(\underset{n-2}{A}\, p^\alpha_{;\alpha} + 2\underset{n-2}{A}{}_{,\alpha} p^\alpha + \underset{n-1}{A}\, p^2) -$$
$$- (\alpha + 2\beta)(\underset{n-2}{A}\, p_{\mu;\nu} + \underset{n-2}{A}{}_{,\mu} p_\nu + \underset{n-2}{A}{}_{,\nu} p_\mu + \underset{n-1}{A}\, p_\mu p_\nu]{}_+z^{n-3} +$$
$$+ \cdots\} + \cdots = \kappa \tau_{\mu\nu}\,{}_+z^{n-1}. \tag{A.14}$$

To discuss the structure of the jumps which are compatible with (A.14) we shall now distinguish the two cases $p^2 = 1$ and $p^2 = 0$ and investigate the equations for $\sigma \equiv \alpha + 3\beta \neq 0$ and $\sigma = 0$ separately.

A.1. Timelike Thin Massive Shells ($p^2 = 1$)

(a) $\sigma \neq 0$

Because of the general covariance of the field equations (A.2) and the Bianchi identity resulting from this, the left-hand sides of (A.14) are not algebraically independent. They satisfy the identity

$$[\alpha \underset{n-2}{A}{}_{\mu\nu} p^2 + (\tfrac{1}{2}\alpha + 2\beta) g_{\mu\nu} \underset{n-2}{A}{}^{\rho}_{\rho} p^2 - (\alpha + 2\beta) \underset{n-2}{A}{}^{\rho}_{\rho} p_\mu p_\nu] p^\nu$$
$$= \alpha \underset{n-2}{A}{}_{\mu\nu} p^\nu - \tfrac{1}{2}\alpha p_\mu \underset{n-2}{A}{}^{\rho}_{\rho} = \alpha(\underset{n}{\gamma}{}_{\mu\nu} p^\nu + \underset{n}{\gamma}{}^{\rho}_{\rho} p_\mu - \underset{n}{\gamma}{}_{\mu\tau} p^\tau -$$
$$- \underset{n}{\gamma}{}_{\nu\tau} p^\nu p^\tau p_\mu - \underset{n}{\gamma}{}^{\rho}_{\rho} p_\mu - \underset{n}{\gamma}{}_{\nu\tau} p^\nu p^\tau p_\mu) = 0. \tag{A.15}$$

Accordingly, the field equations provide only six independent conditions for the jump coefficients $\underset{n}{\gamma}{}_{\mu\nu}$ such that, as in GRT, there exist four solutions of the homogeneous equations which may be produced by the choice of coordinates $x^\nu \to \bar{x}^\nu$:

$$\bar{x}^\nu = x^\nu - \underset{l}{a}{}^{\nu+} z^{l+1}, \; \underset{l}{\bar{\gamma}}{}_{\mu\nu} = \underset{l}{a}{}_\mu p_\nu + \underset{l}{a}{}_\nu p_\mu. \tag{A.16}$$

On the other hand, the Bianchi identity provides also four conditions on the inhomogeneity,

$$\tau_{\mu\nu} p^\nu = 0, \tag{A.17}$$

such that only six of the ten components of the surface layer of density are algebraically independent. Equation (A.17) corresponds to the dynamical equation of GRT and says that the jump surface must be a 'free surface'.

From the trace of (A.14)

$$2 \underset{n-2}{A} l_{\mathrm{P}}^2 (\alpha + 3\beta)_+ z^{n-4} + \cdots = \kappa \tau_+ z^{-1}, \tag{A.18}$$

it is seen that, for $\underset{n-2}{A} \neq 0$, one has

$$n = 3,$$
$$\underset{n-2}{A} = \frac{\kappa\tau}{2 l_{\mathrm{P}}^2 (\alpha + 3\beta)} \tag{A.19}$$

and, using (A.14), one obtains

$$\alpha \underset{n-2}{A}{}_{\mu\nu} = \frac{\kappa}{l_{\mathrm{P}}^2} \hat{\tau}_{\mu\nu}, \tag{A.20}$$

where

$$\hat{\tau}_{\mu\nu} = \tau_{\mu\nu} - \frac{\tau}{2(\alpha + 3\beta)} [(\tfrac{1}{2}\alpha + 2\beta)g_{\mu\nu} - (\alpha + 2\beta)p_\mu p_\nu].$$

For $\underset{n-2}{A} = 0$, there exist only $n = 3$ jumps, too, which follow from (A.20) by assuming $\tau = 0$. Indeed, if one puts $\underset{n-2}{A} = 0$ in (A.18), then the next term $\sim {}_+z^{n-3}$ must be compensated by $\kappa\tau_+z^{-1}$, i.e., one obtains $n = 2$. However, then (A.14) provides

$$\underset{n-2}{A}_{\mu\nu} = 0.$$

This system of linear homogeneous equations does not provide non-trivial solutions for $\underset{n}{\gamma}_{\mu\nu}$ if $\rho^2 \neq 0$.

In GRT one obtains similar equations for the jumps of the first derivatives of $g_{\mu\nu}$ generated by $\kappa\tau_+z^{-1}$, namely

$$\underset{n-2}{A}_{\mu\nu} = \kappa(\tau_{\mu\nu} - \tfrac{1}{2}g_{\mu\nu}\tau). \tag{A.21a}$$

Their solution reads (Lanczos, 1922, 1924; Synge, 1960)

$$\underset{1}{\gamma}_{\mu\nu} = a_\mu p_\nu + a_\nu p_\mu - 2\kappa(\tau_{\mu\nu} - \tfrac{1}{2}g_{\mu\nu}\tau), \tag{A.21b}$$

where a_μ is an arbitrary vector, which can be removed by coordinate transformations which are discontinuous in the second derivatives on $z = 0$.

In analogy to (A.21), we can immediately write down the solution of (A.20):

$$\underset{3}{\gamma}_{\mu\nu} = a_\mu p_\nu + a_\nu p_\mu - \frac{2\kappa}{l_\mathrm{P}^2}\hat{\tau}_{\mu\nu}. \tag{A.22}$$

Here the coordinate-produced jump $a_\mu p_\nu + a_\nu p_\mu$ can be removed by coordinate transformations which are discontinuous in the fourth derivatives on $z = 0$. For Eddington's case $\alpha = -2\beta$ the solution (A.22) fulfils the same coordinate conditions as in GRT, namely

$$(\underset{3}{\gamma}_{\mu\nu} - \tfrac{1}{2}g_{\mu\nu}\underset{3}{\gamma}{}^\rho_\rho)p^\nu = a_\mu. \tag{A.23}$$

It is a typical feature of the jumps given by (A.22), and similar formulae appearing in the following, that the small gravitational constant is compensated by the small Planck length l_P, such that, in contrast to GRT, the first derivatives of the metric are continuous and the jumps of the third derivatives are large. In the quantum case these jumps are connected with the vacuum polarization near the shell. This polarization is essential in the region with a size comparable to or less than the Planckian length l_P. If we take $z \approx l_\mathrm{P}$, then the change of the metric on this length will be given by

$g^+_{\mu\nu} \approx g^-_{\mu\nu} + l_P(-2\kappa)\hat{\tau}_{\mu\nu}$ so that

$$[\partial_p g_{\mu\nu}] \propto \left[\frac{g^+_{\mu\nu} - g^-_{\mu\nu}}{l_P} \right) \propto -2\kappa\hat{\tau}_{\mu\nu}.$$

In this sense, when we are interested in the behaviour of the metric in the spacetime regions larger than l_P, then we can consider it as obeying the usual Einstein jump conditions on the mass shell.

(b) $\sigma = 0$

For $\alpha + 3\beta = 0$ we obtain from (A.1) the trace equations

$$R = -\kappa T^{\mu}_{\mu} \tag{A.24}$$

that is

$$\underset{n-2}{A}{}_{+} z^{n-2} + \underset{n-1}{A}{}_{+} z^{n-1} + \cdots = -\kappa\tau_{+} z^{-1}. \tag{A.25}$$

Accordingly, in the case of the Bach—Weyl equations where only the part $\propto l_P^2$ of the left-hand terms in (A.2) is considered, the trace is identical to zero such that it follows $T^{\mu}_{\mu} = 0$. The pure Bach—Weyl equations are therefore equivalent to nine algebraically independent equations and provide together with the Bianchi identity only five conditions on the $g_{\mu\nu}$. If one includes the Einstein—Hilbert part $G_{\mu\nu}$ in the field equations, the trace equation (A.24) is supplemented and there are again six conditions imposed on the $g_{\mu\nu}$. Consequently now T^{μ}_{μ} need not automatically vanish.

However, it is easily seen that for the case of surface layers of density considered here, the condition $\tau = 0$ must be satisfied. Of course, to show this one has to consider that the arguments referring to the number of functions $g_{\mu\nu}$ and the number of field equations cannot be translated directly to the jump coefficients and the conditions to be fulfilled because (A.11) and (A.14) are differential equations of different order. It depends on the order n of $\underset{n}{\gamma}_{\mu\nu}$ whether the matter term $\tau_{\mu\nu}$ plays a role for $\underset{n}{\gamma}_{\mu\nu}$ in (A.14) or in (A.25).

Let us again distinguish the two cases $\tau \neq 0$ and $\tau = 0$. For $\tau \neq 0$ one obtains from (A.25), $n = 1$ and, as in the case of the pure Bach—Weyl equations, from (A.14), the homogeneous jump condition

$$\underset{n-2}{A}{}_{\mu\nu} = \tfrac{1}{6} g_{\mu\nu} \underset{n-2}{A}{}^{\rho}_{\rho} + \tfrac{1}{3} \underset{n-2}{A}{}^{\rho}_{\rho} p_{\mu} p_{\nu}. \tag{A.26}$$

However, this equation does not provide a non-trivial solution. Indeed, its solution may be written formally as

$$\underset{1}{\gamma}_{\mu\nu} = a_{\mu} p_{\nu} + a_{\nu} p_{\mu} - \tfrac{1}{3} g_{\mu\nu} \underset{n-2}{A}{}^{\rho}_{\rho}. \tag{A.27}$$

This solution is trivial because it can be removed by coordinate transformation together with a conformal transformation. (The situation changes however if the conformal invariance is destroyed by adding the Einstein tensor $G_{\mu\nu}$; cf. Section 6.1.)

For $\tau = 0$ one obtains from (A.25)

$$\underset{n-2}{A} = \underset{n-1}{A} = \cdots = 0 \tag{A.28}$$

and we find, via (A.14)

$$n = 3$$

$$\alpha \underset{n-2}{A}{}_{\mu\nu} = \frac{\kappa}{l_{\mathrm{P}}^2} \tau_{\mu\nu}. \tag{A.29}$$

The solution of (A.29) results from (A.22) if we replace $\hat{\tau}_{\mu\nu}$ by $\tau_{\mu\nu}$. Additionally, one has from (A.26) the condition $\underset{3}{\gamma} = 0$ on $\underset{3}{\gamma}{}_{\mu\nu}$.

A.2. Null Thin Massive Shells ($p^2 = 0$)

(a) $\sigma \neq 0$

The trace of (A.14)

$$2(\alpha + 3\beta)(\underset{n-2}{A} p^{\alpha}_{;\alpha} + 2 \underset{n-2}{A}{}_{,\alpha} p^{\alpha})_{+} z^{n-3} + \cdots = \kappa \tau_{+} z^{-1} \tag{A.30}$$

shows that one has to distinguish again the cases $\underset{n-2}{A} = 0$ and $\underset{n-2}{A} \neq 0$.

If $\underset{n-2}{A} = 0$, then, by virtue of (A.30), one obtains $n = 1$ or $n \geqslant 2$ (with $\tau = 0$). It is seen from (A.14), firstly, that $n > 2$ must be excluded because $\tau_{\mu\nu} \neq 0$ is assumed and, secondly that the following conditions have to be satisfied:

$$n = 1,$$

$$\alpha(2 \underset{n-2}{A}{}_{\mu\nu;\alpha} p^{\alpha} + \underset{n-2}{A}{}_{\mu\nu} p^{\alpha}_{;\alpha}) - (\alpha + 2\beta) \underset{n-1}{A} p_{\mu} p_{\nu} = 0 \tag{A.31a}$$

or

$$n = 2,$$

$$\tau_{\mu\nu} = \frac{l_{\mathrm{P}}^2}{\kappa} [\alpha(2 \underset{n-2}{A}{}_{\mu\nu;\alpha} p^{\alpha} + \underset{n-2}{A}{}_{\mu\nu} p^{\alpha}_{;\alpha}) - (\alpha + 2\beta) \underset{n-1}{A} p_{\mu} p_{\nu}]. \tag{A.31b}$$

Assuming $\underset{n-2}{A} \neq 0$, (A.30) leads to $n = 1$, $n = 2$, or $n \geqslant 3$ (with $\tau = 0$), and (A.14) demonstrates that $n > 3$ and $n = 1, 2$ are not compatible with the field equations because of $\tau_{\mu\nu} \neq 0$ and $\underset{n-2}{A} \neq 0$, respectively. For $n = 3$,

(A.14) furnishes

$$\tau_{\mu\nu} = -\frac{\alpha + 2\beta}{\kappa} l_{\mathrm{P}}^2 \underset{n-2}{A} p_\mu p_\nu. \tag{A.32}$$

(b) $\sigma = 0$

Regarding the equation (A.25) one must again exclude the case $\tau \neq 0$ from consideration, since (A.25) leads in this case to the requirements $n = 1$ and $\underset{n-2}{A} = \tau \neq 0$, which contradict (A.14). Consequently, one has to assume $\tau = 0$ and, because of (A.21), $\underset{n-2}{A} = \underset{n-1}{A} = \cdots = 0$.

Repeating the considerations given above for $\sigma \neq 0$ regarding the constraint $\underset{n-2}{A} = 0$, we obtain the two jumps (A.31a) and (A.31b), where $\underset{n-2}{A}$ must be now assumed to be zero.

Massive shell models described by surface layers of density are of course connected with jumps of the derivatives of the metric. Shells moving with a velocity smaller than the velocity of light always cause, independent of the α and β values, discontinuities in the third derivatives ($n = 3$). Their structure is determined by the matter density $\tau_{\mu\nu}$ via (A.22). This corresponds completely to GRT, where the same applies to $n = 1$ discontinuities. The two cases $\sigma = 0$ and $\sigma \neq 0$ differ only in that for $\sigma = 0$ the trace τ has to be equal to zero.

For shells moving with the velocity of light, there exists a greater variety of possible jumps because the shell motion can be accompanied by free gravitational shock waves of different order. Here one finds typical differences between the $\sigma = 0$ and $\sigma \neq 0$ cases.

The discontinuities on null surfaces induced by $\tau_{\mu\nu}$ are, for $\sigma \equiv (\alpha + 3\beta) \neq 0$, in general (i.e., if $\underset{n-2}{A} \neq 0$) of third order and, for $\sigma = 0$, of second order. Only in the $\sigma \neq 0$ case is $\tau_{\mu\nu}$ necessarily of the form $\tau_{\mu\nu} \propto p_\mu p_\nu$. It is interesting to note that for $\alpha = -2\beta$ the formulae for $\sigma \neq 0$ reduce to the $\sigma = 0$ relations if $p_\rho p^\rho = 0$. The $\alpha = -2\beta$ case considered by Eddington long ago is to a certain extent analogous to the $\alpha = -3\beta$ case. The similarity has already been observed in the case of free gravitational shock waves with $n = 2$ discussed in von Borzeszkowski *et al.*, 1978.

Besides the jumps of second and third order generated by $\tau_{\mu\nu}$ on null surfaces, there can also appear jumps of the first and second derivatives, corresponding to free gravitational shock waves, while the $n = 3$ jumps are determined by $\tau_{\mu\nu}$ via (A.32). There are no free $n = 3$ shock waves accompanying the motion of the matter shell. For free gravitational shock waves the algebraic structure of the coefficients $\underset{1}{\gamma}_{\mu\nu}$ and $\underset{2}{\gamma}_{\mu\nu}$ is not determined by the field equations. Equations of the type (A.31a) provide only conditions on the propagation along the bicharacteristics $z = 0$.

Finally it should be mentioned that our considerations demonstrate that, for $\alpha = -3\beta$ (i.e., if the higher-order part of the field equations is conform-invariant), the trace τ of the matter density has necessarily to be equal to zero. One finds the same result for $\alpha = -2\beta$.

Appendix B

On the Physical Meaning of Planck's 'Natural Units'*

Max Planck (1899)

Es dürfte nicht ohne Interesse sein zu bemerken, daß mit Zuhülfenahme der beiden in dem Ausdruck der Strahlungsentropie auftretenden Constanten die Möglichkeit gegeben ist, Einheiten für Länge, Masse, Zeit und Temperatur aufzustellen, welche unabhängig von speciellen Körpern oder Substanzen, ihre Bedeutung für alle Zeiten und für alle, auch ausserirdische und aussermenschliche Culturen notwendig behalten und welche daher als "natürliche Maasseinheiten" bezeichnet werden.

Diese Größen behalten ihre natürliche Bedeutung so lange bei, als die Gesetze der Gravitation, der Lichtfortpflanzung im Vacuum und die beiden Hauptsätze der Wärmetheorie in Gültigkeit bleiben, sie müssen also, von den verschiedensten Intelligenzen nach den verschiedensten Methoden gemessen, sich immer wieder als die nämlichen ergeben.

I. Already before the final formulation of quantum theory, Planck (1899) had derived the existence of a new universal constant from Wien's law, namely, the action quantum $\hbar$. Planck remarked that, using $\hbar$ and the two other universal constants, the gravitational constant G and the velocity of light *in vacuo* c, it was possible to define natural units for mass, length, and time; he said that these units had a meaning always and everywhere in the universe.**

Speaking in the language of modern physics, Planck's system of natural units is based on the fundamental constant of quantum theory (Planck's action quantum $\hbar$), the fundamental constant of the theory of special relativity (the velocity of light c), and the fundamental constant of the theory

* Cf. for this Appendix, Treder (1979). See also the discussion between Källen (1965) and Treder (1965).

** Planck took over the section on the natural units from his 1899 paper in his review article *Über irreversible Strahlungsvorgänge* (1900b). There is a section on natural units in the first three editions and also in Planck's famous book *Theorie der Wärmestrahlung* (1906). In the fourth edition Planck omitted this section as 'dispensible'.

of general relativity (the gravitational constant G). Therefore, Planck's ideas concerning the physical meaning of his elementary units imply the conception that all physics may be reduced to quantum theory and the theories of special and general relativity. This means that Planck's ideas imply the conception that gravity is not only a universal but also a fundamental interaction and that the gravitational field also constitutes elementary particles.

In this sense, Planck's units involve Einstein's particle scheme (Einstein, 1919, 1936; Einstein and Rosen, 1935), namely, a representation of elementary particles as self-consistent solutions of the relativistic equations of gravity. According to Planck and Einstein, the masses of the elementary particles were then given by their gravitational interaction, i.e., the energy of their own gravitational fields. Thus, Planck's units lead from the Abraham–Lorentz conception, according to which the masses of particles are attributed to their electromagnetic field energies, to Einstein's general-relativistic particle problem. Planck's idea of introducing the action quantum $\hbar$ as one of the fundamental constants implies also the idea that Einstein's particle problem cannot be solved on the level of the classical theory but only within the framework of a quantum theory of the gravitational field, i.e., of a quantum geometrodynamics.

Probably Eddington (1918) considered Planck's natural units as an ansatz for unifying physics on the basis of a relativistic theory of gravity, too. On the other hand, in a review of Planck's and Eddington's conceptions, Bridgman (1922) criticized the physical meaning of Planck's units because Planck could not show any essential connection between the theories leading to the constants $\hbar$, c, and G.

To our minds, quantum geometrodynamics and the quantum theory for gravitational fields just show the connection required by Bridgman. The revival of quantum geometrodynamics and of a quantum theory of the spacetime metrics also renewed interest in Planck's natural units (Wheeler, 1962, 1968).

If physics is to reduce to Planck's constants $\hbar$, c, and G, then this means that the elementary charge referred to these fundamental constants must be of order 'one'. Thus, Planck's elementary charge is given by

$$[q]^2 = \rho_P^2 = \hbar c \approx 10^{-16}\,\mathrm{g\,cm^3\,s^{-2}},$$

$$[q] = \rho_P = (\hbar c)^{1/2} \approx 10^{-8}\,\mathrm{cgs}. \tag{B.1}$$

According to Planck's ansatz, gravity is the fundamental interaction; therefore, it follows from the gravitational law that

$$q_P^2 = G m_P^2,$$

$$q_P = \sqrt{G}\, m_P \tag{B.2}$$

and hence we obtain Planck's elementary mass:

$$[m] = m_P = \left(\frac{\hbar c}{G}\right)^{1/2} \approx 10^{-5}\ \mathrm{g}. \tag{B.3}$$

The field energy of a mass m_P defined by Einstein has a positive sign in the theory of general relativity; this renders it possible to formulate Einstein's particle problem (Einstein, 1919). According to the Einsteinian principles of the equivalence of inertia and gravity, and the equivalence of mass and energy, the gravitational self-energy of a particle belonging to the mass (B.3) must be equal to its relative energy $m_P c^2$ if the particle energy is essentially of gravitational origin. This leads to the condition (up to a factor of order one, which characterizes the special 'model of the particle')

$$\frac{G m_P^2}{l_P} = m_P c^2 \tag{B.4}$$

from which, on using relation (B.3), the effective radius of the particle follows:

$$[l] = l_P = \frac{G m_P}{c^2} = \left(\frac{\hbar G}{c^3}\right)^{1/2} \approx 10^{-33}\ \mathrm{cm}. \tag{B.5}$$

The elementary time

$$[t] = t_P = \frac{l_P}{c} = \left(\frac{\hbar G}{c^5}\right)^{1/2} \approx 10^{-43}\ \mathrm{s} \tag{B.6}$$

corresponds to the elementary length (B.5).

It is essential for the following that the inertial and thus the gravitational masses of a particle are not invariants but proportional to the fourth, timelike component P_0 of the relativistic energy—momentum vector P. Relations (B.2) and (B.3) hold in a special-relativistic system of reference not being the rest system of the particle.

II. To understand the physical meaning of Planck's units, one must recall the fact that relations (B.1) and (B.2) say that, in Planck's system of reference, gravity is a strong interaction; for a particle that rests in the laboratory (e.g., nucleons) gravity is an ultraweak force. The conception of quantum field theory about the nature of elementary particles leads to unitary field theories of the type of Heisenberg, Pauli, and Ivanenko.

Like Planck, Heisenberg starts with three fundamental constants, too. But, these are the three fundamental constants $\hbar$, c, and the rest mass of a nucleon. The coupling constant of a unitary field theory of the Heisenberg—

Ivanenko type is also determined by Planck's elementary charge,

$$[q] = g = (\hbar c)^{1/2} \approx 10^{-8}\,\text{cgs}. \tag{B.7}$$

In (B.3) g is the coupling constant of the strong interaction and plays a role that justifies the notation 'strong nucleon charge' (Ivanenko and Sokolov, 1953); as the electric charge e (the coupling constant of the electromagnetic interaction) g is a Lorentz-invariant scalar too (Heisenberg, 1967).

The idea that, in the sense of Abraham's particle problem, the rest energy mc^2 of a nucleon is determined by the self-energy of its strong interaction, defines Heisenberg's elementary length l in the rest system as an effective radius of the nucleon (assuming the numerical factor for the 'particle model' and the special type of coupling to be equal to (B.1)):

$$\frac{g^2}{l} = mc^2 \approx 10^{-3}\,\text{g cm}^2\,\text{s}^{-2}. \tag{B.8}$$

Inserting (B.7), the nucleon radius becomes equal to the Compton wavelength of nucleons:

$$[l] = l = \frac{g^2}{mc^2} = \frac{\hbar c}{mc^2} = \frac{\hbar}{mc} \approx 10^{-13}\,\text{cm}. \tag{B.9}$$

Equation (B.9) is the elementary length of March, Heisenberg, and Ivanenko.

III. Let now the nucleon be moving ultrarelativistically (i.e., with a velocity $v \approx c$) with reference to the system of the laboratory. Denoting the corresponding Lorentz factor by

$$\gamma = (1 - v^2/c^2)^{-1/2} \tag{B.10}$$

one obtains the de Broglie wavelengths in the form of the Lorentz-contracted Compton wavelength

$$L' = \frac{\hbar c}{E'} = \frac{1}{\gamma}\, l = \frac{\hbar}{\gamma mc} \tag{B.11}$$

and the inertial mass of the ultrarelativistic nucleon is given by

$$m' = \gamma m = \frac{E'}{c^2}. \tag{B.12}$$

Simultaneously the effective radius of the nucleon is contracted,

$$l' = \frac{l}{\gamma} = L'. \tag{B.13}$$

Let us assume now that the velocity of the ultrarelativistic nucleon is, with reference to the laboratory, so high that the contracted radius of the nucleon

(B.13) corresponds to the gravitational radius

$$\frac{Gm'}{c^2} = \frac{G\hbar}{L'c^3} \tag{B.14}$$

of the inertial mass (B.12). Therefore, it reads

$$\frac{\gamma Gm}{c^2} = \frac{Gm'}{c^2} = \frac{\hbar}{m'c} = \frac{\hbar}{\gamma mc}. \tag{B.15}$$

Equation (B.15) determines the Lorentz factor

$$\gamma = \gamma^* = \left(\frac{\hbar c}{G}\right)^{1/2} \frac{1}{m} \approx 10^{20}. \tag{B.16}$$

Using this relation to transform Heisenberg's quantities and m from the rest system of the particle to the reference system of the laboratory, one obtains Planck's quantities l_P and m_P. Simultaneously the gravitational coupling becomes strong (Treder, 1975b):

$$\gamma^{*2} Gm^2 = Gm_P^2 = \frac{\hbar^2 G}{L^{*2} c^2} = \hbar c.$$

Thus, Planck's elementary quantities follow from Heisenberg's quantities by a Lorentz transformation with the Lorentz factor (B.16). By this Heisenberg's elementary length, i.e., the Compton wavelength $\hbar/mc$ of the nucleons, contracts to de Broglie's wavelength $L^* = \hbar c/E^*$ of ultrarelativistic particles with the inertial mass $m_P = \gamma m = m^*$.

IV. For scalar charges $[e]$ the Lorentz contraction of the radii of particles $[l]$ and the increase of its self-energies $\sim [e]^2[l]^{-1}$ resulting from this just give the relativistic increase of the mass, e.g.,

$$\frac{g^2}{l'} = \frac{\gamma g^2}{l} = m'c^2 = \gamma mc^2. \tag{B.17}$$

As one knows, according to Lorentz this is the content of the special-relativistic theory of field masses.

However, in the theory of general relativity there is quite another relation between the gravitational field mass $\sim Gm'^2/l'c^2$ and the inertial mass m' of a moving particle because the 'gravitational charge' itself has to be Lorentz-transformed:

$$[q'] = \sqrt{G}\, m' = \gamma \sqrt{G}\, m. \tag{B.18}$$

Hence it follows that the gravitational self-energy increases more rapidly than the inertial mass (the relative energy) of particles, which is determined by the

Lorentz—Einsein formula:

$$\frac{Gm'^2}{l'} = \frac{G\gamma^3 m^2}{l}. \tag{B.19}$$

Therefore, energy and momentum of the stationary field of a mass cannot form a stable particle; for stable particles the energy—momentum must form a special-relativistic 4-vector p_μ (Ivanenko and Solokov, 1953). Equation (B.19) says that there are rapidly increasing pressures and tensions in the interior of particles. According to Einstein's original idea, these pressures and tensions are of an electromagnetic nature.

The Lorentz relation (B.17) connecting field and inertial masses is satisfied for gravitational fields in a system of reference where the Lorentz factor has the value $\gamma = \gamma^*$ and the inertial mass $m' = \gamma E/c^2 = E'/c^2$ and the particle radius $l' = \gamma^{-1} l$, respectively, are just equal to Planck's quantities m_P and l_P, respectively:

$$\gamma^2 = \gamma^{*2} = \frac{lc^2}{Gm} = \frac{\hbar c}{G}\frac{1}{m^2}. \tag{B.20}$$

The gravitational potential $\phi = Gm/r$ is itself transformed as the time—time component of the metric tensor $g_{\mu\nu}$. According to the theory of general relativity, the gravitational field of a point particle is given by the Schwarzschild metric.* In the rest system of the particle, this solution reads in the linear approximation:

$$\begin{aligned} &g_{00} - 1 = h_{00} = -2\phi/c^2, \\ &g_{ik} = -\delta_{ik} + h_{ik} = -\delta_{ik}(1 + 2\phi/c^2), \\ &i, k = 1, 2, 3, x^1 = X, x^2 = Y, x^3 = Z. \end{aligned} \tag{B.21}$$

If now the particle is moving with velocity v in the X direction, the corresponding Lorentz transformation gives the time—time component

$$\begin{aligned} h'_{00} &= \alpha_0^0 \alpha_0^0 h_{00} + \alpha_0^1 \alpha_0^1 h_{11} = \gamma^2 \left(h_{00} + \frac{v^2}{c^2} h_{11} \right) \\ &= -\frac{(1 + v^2/c^2)}{(1 - v^2/c^2)} \frac{2\phi}{c^2} = \frac{2\phi'(\gamma^2)}{c^2}. \end{aligned} \tag{B.22}$$

* The one-particle vacuum solution (the gravitational potential of a mass point) of Einstein's gravitation equations, i.e., the Schwarzschild metric (Einstein, 1969), is no more able to describe a self-consistent particle than the Coulomb field to describe a self-consistent electron in Maxwell's theory. Therefore, the Einstein energy—momentum complex $\int \sqrt{-g}\,(T_\mu^0 + \mathfrak{t}_\mu^0)\,\mathrm{d}^3x$ of a mass point does not form an energy—momentum vector p_μ and the gravitational field must be renormalized (e.g., by a cut-off formalism with an elementary length and by 'compensating fields').

Hence it follows for the gravitational potential of an ultrarelativistically moving particle, the approximate expression (with $v \approx c$)

$$\phi' = \frac{Gm'}{\gamma'} \approx \frac{2Gm}{[(1 - v^2/c^2)X^2 + Y^2 + Z^2]^{1/2}(1 - v^2/c^2)^{1/2}}$$

$$\approx \frac{2Gm\gamma}{[(X^2/\gamma^2) + Y^2 + Z^2]^{1/2}} \,. \tag{B.23}$$

Indeed, going to a moving system of reference, the gravitational mass m becomes $m' = \gamma m$, while the distances are Lorentz-contracted in the direction of motion $X' = \gamma^{-1}X$.

A particle having effective radius l and producing the maximal gravitational potential $\phi(l) = Gm/l$ in the rest system, accordingly has the maximal gravitational potential

$$\phi'(l') = \frac{2Gm'}{l'} = \frac{2\gamma^2 Gm}{l} \tag{B.24}$$

in the moving system of reference (in the direction of motion). If the Lorentz factor runs to (B.20), then the maximal potential is just of the order of $2c^2$:

$$\phi'(l') = \phi^*(l^*) = \frac{2Gm_P}{l_P} = \frac{2G(\hbar c/G)^{1/2}}{(\hbar G/c^3)^{1/2}} = 2c^2 \tag{B.25}$$

and, for a Lorentz factor $\gamma = \bar{\gamma} > \gamma^*$, i.e., if the effective particle radius contracts below Planck's length l_P, the maximal gravitational potential $\overline{\phi}(l)$ is greater than $2c^2$, such that we obtain

$$\bar{g}_{00}(\bar{l}) \approx 1 - \frac{2\overline{\phi}(\bar{l})}{c^2} \approx 1 - 4\frac{Gm}{c^2 l}\bar{\gamma}^2 < 0. \tag{B.26}$$

Then there exists a region in the vicinity of the particle where $\bar{g}_{00} < 0$, i.e., where the timelike world quantities change their signature (cf. Treder, 1975b). Therefore, the often-discussed anomalies of causality that are connected with the Schwarzschild surface $g_{00} = 0$ appear near the particle.*

Now the particle lies within a black hole, whose linear dimension is of the order of the Einsteinian gravitational radius

$$G\frac{\bar{m}}{c^2} = \bar{\gamma}\frac{Gm}{c^2} > \bar{l} = \frac{\hbar}{\bar{m}c} = \frac{\hbar}{\bar{\gamma}mc}$$

* The Schwarzschild surface $g_{00} = 0$ implies, as a consequence of the change of signature in the region enclosed by the Schwarzschild surface, the destruction of the causal connection between the interior $g_{00} < 0$ and the exterior $g_{00} > 0$ of the space. In 1935 Einstein had already mentioned the meaning of these signature anomalies for the solution of the particle problem (Einstein and Rosen, 1935).

of the ultrarelativistic particle. Assuming $m' = m_P$, one obtains

$$\frac{Gm_P}{c^2} = \frac{\hbar}{m_P c} = l_P$$

and the Einsteinian gravitational radius (i.e., the Schwarzschild radius of the black hole), the classical radius of particles, and the de Broglie wavelength are of equal order.

V. At the boundary of the black hole a total Einstein shift of frequencies occurs; all processes slow down infinitely with reference to the time of the laboratory:

$$\tau = \frac{\sqrt{\bar{g}_{00}}\,t}{\gamma} \to 0, \; \bar{g}_{00} \to 0. \tag{B.27}$$

It is impossible to obtain information from the black hole, i.e., from regions where

$$\bar{g}_{00} \leqslant 0, \phi \gtrsim \tfrac{1}{2}c^2. \tag{B.28}$$

Since now measurements of differences of lengths ΔL must be done with particles as standards whose effective radius and de Broglie wavelength $L' = \hbar c/E'$ are less than ΔL, Planck's length l_P (for which the de Broglie wavelength is equal to Einstein's gravitational radius) is actually the shortest measurable length, according to the unified quantum and relativity theories. It is true, for $\bar{E} > E^*$, that the matter wavelength $\bar{L}$ is smaller than l_P: $\bar{L} = \hbar c/\bar{E} < l_P$. But the gravitational radius $G\bar{E}/c^4$ and, therefore, the effective radius of particles is greater than l_P. Being the equivalence mass of an elementary particle that has the ultrarelativistic energy $E^* = \gamma^* mc^2$, Planck's mass m_P determines the maximal energy of an elementary particle too. Then, for $\bar{E} > E^*$, the particle is not connected causally with its vicinity (with the laboratory). The strong gravity of the particle eliminates it from physical space. This is the physical meaning of the new uncertainty relation discussed above.

Evidently here this property is relative. Looking at the laboratory (and the whole cosmos) from the rest system of the nucleon, they form black holes. The decisive point is that, for a Lorentz factor $\bar{\gamma} > \gamma^* = (\hbar c/G)^{1/2}(1/m)$, the laboratory and the nucleons are not connected causally. Indeed, the laboratory (or the cosmos) is formed from the same nucleons with the rest mass m as the particle in its interior. We can model our problem by two nucleons whose relative velocity is given by the Lorentz factor $\bar{\gamma} > \gamma^* = m^*/m$. Evidently here there exists complete symmetry between the assertions. Nucleon I is moving with γ in the rest system of nucleon II, and nucleon II is moving with $\bar{\gamma}$ in the rest system of nucleon I. In both cases the causal connection between both the particles is destroyed.

VI. In general, the occurrence of Schwarzschild horizon, being here only the result of a Lorentz transformation of the metric according to (B.27), depends on the relative velocity v of the nucleon with regard to the reference system.

Let two (ultrarelativistic) particles I and II have velocities v_{I} and v_{II} with reference to the laboratory and, therefore, the Lorentz factors

$$\gamma_{\mathrm{I}} = \left(1 - \frac{v_{\mathrm{I}}^2}{c^2}\right)^{-1/2} \qquad \gamma_{\mathrm{II}} = \left(1 - \frac{v_{\mathrm{II}}^2}{c^2}\right)^{-1/2}$$

respectively, in the system of the laboratory. Then their relative velocity $v_{\mathrm{I\,II}}$ is given by Einstein's theorem for the combination of two boosts in the same direction

$$v_{\mathrm{I\,II}} = \frac{v_{\mathrm{I}} + v_{\mathrm{II}}}{1 + (v_{\mathrm{I}} v_{\mathrm{II}}/c^2)}\,.$$

Their relative Lorentz factor $\gamma_{\mathrm{I\,II}} = [1 - v_{\mathrm{I\,II}}^2/c^2]^{-1/2}$ reads

$$\gamma_{\mathrm{I\,II}} = \left(1 + \frac{v_{\mathrm{I}} v_{\mathrm{II}}}{c^2}\right)\gamma_{\mathrm{I}}\gamma_{\mathrm{II}}. \tag{B.29}$$

For $v_{\mathrm{I}} \approx v_{\mathrm{II}} \approx \pm c$ it is bilinear in γ_{I} and γ_{II}:

$$\gamma_{\mathrm{I\,II}} \approx 2\gamma_{\mathrm{I}}\gamma_{\mathrm{II}} \tag{B.29a}$$

while, for $v_{\mathrm{I}} \approx -v_{\mathrm{II}}$, the relative velocity of the particles is approximately vanishing and $\gamma_{\mathrm{I\,II}} \to 1$. Expression (B.29a) gives the transformation formula for the time component p_0 of the energy—momentum vector p_μ of an ultrarelativistic particle with $p_\mu p^\mu \approx 0$:

$$E_{\mathrm{I\,II}} \approx \gamma_{\mathrm{I}}\gamma_{\mathrm{II}} mc^2 \approx 2\gamma_{\mathrm{I}} E_{\mathrm{II}} \approx 2\gamma_{\mathrm{II}} E_{\mathrm{I}} \approx 2\gamma_{\mathrm{I}}\gamma_{\mathrm{II}} mc^2 \tag{B.30}$$

(with $v_{\mathrm{I}} \approx v_{\mathrm{II}} \approx \pm c$).* If $\gamma_{\mathrm{I\,II}} \gtrsim \gamma^*$, then there exists no causal connection between particles I and II; the particles behave toward each other as if they were black holes.

Thus, the Planckian Lorentz factor γ^* leads to an effective maximum velocity for a system of particles with a rest mass m that is less than the light velocity c. Relative velocities between two particles $v_{\mathrm{I\,II}}$ for which the Lorentz factor is greater than the square of $\sqrt{2}\,\gamma^*$,

$$\gamma_{\mathrm{I\,II}} > 2\gamma^{*2} = 2\,\frac{\hbar c}{G}\,\frac{1}{m^2}$$

* Equation (B.30) is the relation between energies in the rest and the laboratory systems, which is well known from high-energy physics.

cannot occur at all.* Indeed, this could only occur if at least one of the particles had a relative velocity v with reference to the laboratory corresponding to $\gamma > \gamma^*$. But in laboratories such particles cannot be produced or detected because they are outside all causal and information connections between particles and the laboratory.

The role of γ^* as a Lorentz factor for a maximal velocity of a particle of rest mass m with reference to the laboratory system means the dependence of the laboratory velocities really attainable on the kind of particles. Of course, the photons, which have vanishing rest mass ($m \to 0$), always have a Lorentz factor $\gamma \to \infty$ with reference to all particles and, therefore, in all laboratories. But physical laboratories consist of particles with a non-vanishing rest mass m, and the attainable maximum velocity is given by $\gamma^* = (\hbar c/G)^{1/2}(1/m)$, as a reciprocal function of m. On the other hand, the attainable relative maximum mass m_P and the corresponding effective radius of the particle, i.e., the gravitational radius Gm_P/c^2, are equal to Planck's universal quantities $(\hbar c/G)^{1/2}$ and $(\hbar G/c^3)^{1/2}$, respectively, independent of the kind of particles.

* This means that there are no relative velocities $\bar{v}_{\mathrm{I\,II}}$ between two particles for which the effective radii of particles $\bar{l}$ are contracted to the Einstein gravitational radius Gm/c^2 of the particles in the rest system, because then the Lorentz factor $\gamma_{\mathrm{I\,II}}$ had to be so great that

$$\bar{l} \approx \frac{Gm}{c^2} \approx \frac{1}{\gamma}\, l \approx \frac{\hbar}{mc}\, \frac{1}{\gamma}.$$

This leads to

$$\bar{\gamma} \approx \frac{\hbar c}{G}\, \frac{1}{m^2} \approx \gamma^{*2}.$$

References

Adler, S. L.: 1982, *Rev. Mod. Phys.* **54**, 729.

Asthekar, A.: 1983, 'Quantum Gravity', in: B. Bertotti, F. de Felice, and A. Pascolini (eds.), *Proceedings of GR 10*, D. Reidel, Dordrecht, Boston, Lancaster.

Bach, R.: 1921, *Math. Zeitschr.* **9**, 110.

Barut, A. O.: 1974, *Phys. Rev.* **10**, 3335, and reference therein.

Barut, A. O.: 1979, 'Nonlinear Problems in Classical and Quantum Electrodynamics', in: A. F. Rañada (ed.), *Nonlinear Problem in Theoretical Physics*, Lecture Notes in Physics, Springer, Berlin, Heidelberg, New York.

Bergmann, P. G.: 1959, 'General Relativity', in: *Handbuch der Physik*, Springer, Heidelberg.

Bergmann, P. G.: 1979, 'Unitary Field Theories, Yesterday, Today, Tomorrow', in: H.-J. Treder (ed.), *Einstein-Centenarium*, Akademie-Verlag, Berlin.

Birrell, N. D., and P. C. W. Davies: 1982, *Quantum Fields in Curved Space*, Cambridge, U.P.

Bleyer, U., and H.-H. v. Borzeszkowski: 1984, *Ann. Phys. Leipzig* **41**, 241.

Bohr, N.: 1957, *Atomphysik und menschliche Erkenntnis*, Vieweg, Braunschweig, Vol. I.

Bohr, N.: 1966, *Atomphysik und menschliche Erkenntnis*, Vieweg, Braunschweig, Vol. II.

Bohr, N., A. Kramers, and Y. C. Slater: 1924, *Phil. Mag.* **47**, 785.

Bohr, N.: 1949, Discussion with Einstein; in: P. A. Schilpp (ed.), *Albert Einstein: Philosopher — Scientist*, Library of Living Philosophers, New York.

Bohr, N., and L. Rosenfeld: 1933, Zur Frage der Meßbarkeit der elektromagnetischen Feldgrößen (On the question of measurability of electromagnetical field quantities), *Det. Kgl. Danske Videnskabernes Selskab., Math.-fys. Medd.* XII, No. 8, Levin & Munksgaard, Kopenhagen; cf. also in: R. S. Cohen and J. J. Stachel (eds.), *Selected Papers of Léon Rosenfeld*, Boston Studies of the Philosophy of Science, Vol. XXI, D. Reidel, Dordrecht, Boston, London (1979).

Bopp, F.: 1940, *Ann. Phys. Leipzig* **38**, 345.

Born, M.: 1934, *Proc. Roy. Soc. London* **143**, 1410.

Born, M., and L. Infeld: 1934, *Proc. Roy. Soc. London* **A144**, 425.

v. Borzeszkowski, H.-H.: 1981, *Ann. Phys. Leipzig* **38**, 231.

v. Borzeszkowski, H.-H.: 1982, *Found. Phys.* **12**, 633.

v. Borzeszkowski, H.-H.: 1984, 'On High-Frequency Background Quantization of Gravity', in: *Quantum Gravity*, M. A. Markov and P. C. West (eds.), Plenum, New York and London.

v. Borzeszkowski, H.-H.: 1986, *Ann. Phys. Leipzig* **43**, 170.

v. Borzeszkowski, H.-H.: 1985b, 'The Concept of Gravitons and the Measurement of Effects of Quantum Gravity', in: M. A. Markov, V. A. Berezin, and V. P. Frolov (eds.), *Quantum Gravity, Proceedings of the Third Seminar on Quantum Gravity, Moscow 1984*, World Scientific Publishing Co., Singapore.

v. Borzeszkowski, H.-H., and V. P. Frolov: 1980, *Ann. Phys. Leipzig* **37**, 285.

v. Borzeszkowski, H.-H., and H.-J. Treder: 1982a, *Found. Phys.* **12**, 413.

v. Borzeszkowski, H.-H., and H.-J. Treder: 1982b, *Found. Phys.* **12**, 1113; cf. also, 'Quantum Theory and Einstein's General Relativity', in A. O. Barut, A. van der Merwe, and J.-P. Vigier (eds.), *Quantum, Space and Time — The Quest Continues*, Cambridge U.P., London, Cambridge, 1985 (Cambridge Monographs on Physics).

v. Borzeszkowski, H.-H., and H.-J. Treder: 1983a, *Ann. Phys. Leipzig* **40**, 287.

v. Borzeszkowski, H.-H., and H.-J. Treder: 1983b, *Ann. Phys. Leipzig* **40**, 298.

v. Borzeszkowski, H.-H., H.-J. Treder, and W. Yourgrau: 1978, *Ann. Phys. Leipzig* **35**, 471.

Bothe, W., and H. Geiger: 1924, *Z. Phys.* **26**, 44.

Braginsky, V. B., Yu. I. Vorontsov, and F. Ya. Khalili: 1977, *Zhurn. Eksp. Teor. Fiz.* **73**, 1340; Sov. Phys. JETP **46**, 705.
Braginsky, V. B., Yu. I. Vorontsov, and F. Ya. Khalili: 1978, *Pisma ZhETF* **27**, 296; *Sov. Phys. JETP Letters* **27**, 277.
Bridgman, P. W.: 1922, *Dimensional Analysis*, Yale U.P., New Haven, Connecticut.
Brill, D., and J. B. Hartle: 1964, *Phys. Rev.* **B135**, 271.
Buchdahl, H.: 1962, *Nuovo Cimento* **123**, 141.
Capper, D. M., and M. J. Duff: 1974, *Nucl. Phys.* **B82**, 147.
Caves, C. W., R. W. Drever, V. Sandberg, K. S. Thorne, and M. Zimmermann: 1978, *Phys. Rev. Lett.* **40**, 667; 1980, *Rev. Mod. Phys.* **52**, 342.
Choquet-Bruhat, Y.: 1969, *Comm. Math. Phys.* **12**, 16.
Collada, R., A. Overhauser, and S. Werner: 1975, *Phys. Rev. Lett.* **34**, 1472.
Compton, H. A.: 1923, *Phys. Rev.* **21**, 483.
Compton, H. A., and A. W. Simon: 1925, *Phys. Rev.* **26**, 889.
Costa de Beauregard, O.: 1985, *Found. Phys.* (in press).
Debye, P.: 1923, *Phys. Zeitschr.* **24**, 161.
Deser, S., and P. van Nieuwenhuizen: 1974, *Phys. Rev.* **D10**, 401.
DeWitt, B.: 1962, 'The Quantization of Geometry', in: L. Witten (ed.), *Gravitation*, John Wiley and Sons, New York and London.
DeWitt, B.: 1964, 'Dynamical Theory of Groups and Fields', in: C. DeWitt and B. DeWitt (eds.), *Relativity, Groups and Topology*, Blackie and Sons, London and Glasgow.
DeWitt, B. S.: 1967, *Phys. Rev.* **162**, 1195, 1239.
DeWitt, B. S.: 1975, *Phys. Rep.* **19C**, 295.
Dirac, P. A. M.: 1938, *Proc. Roy. Soc. London* **A167**, 148.
Dodonov, V. V., V. I. Man'ko, and V. N. Rudenko: 1983, *Found. Phys.* **13**, 607.
Duff, M.: 1975, 'Covariant Quantization', in: C. J. Isham, R. Penrose, and D. W. Sciama (eds.), *Quantum Gravity*, Clarendon, Oxford.
Eddington, A. S.: 1918, 'Report on Gravitation', *Physical Society, London.*
Eddington, A. S.: 1953, *Fundamental Theory*, Cambridge U.P.
Einstein, A.: 1905, *Ann. Phys. Leipzig* **17**, 132.
Einstein, A.: 1911, *Ann. Phys. Leipzig* **35**, 679.
Einstein, A.: 1919, *Sitzungsber. Preuss. Akad. Wiss.*, p. 349.
Einstein, A.: 1921, *Sitzungsber. Preuss. Akad. Wiss.*, p. 261.
Einstein, A.: 1916, *Die Grundlagen der allgemeinen Relativitätstheorie*, J. A. Barth, Leipzig.
Einstein, A.: 1936, *J. Franklin Inst.* **3**, 221.
Einstein, A.: 1949, 'Remarks', in: P. A. Schilpp (ed.), *Albert Einstein: Philosopher — Scientist*, Open Court, La Salle, Illinois.
Einstein, A., H. A. Lorentz, A. Minkowski, and H. Weyl: 1922, *Das Relativitätsprinzip*, 4th edn., Leipzig.
Einstein, A., and N. Rosen: 1935, *Phys. Rev.* **48**, 73.
Euler, H.: 1936, *Ann. Phys. Leipzig* **26**, 298.
Euler, H., and B. Kockel: 1935, *Naturwiss.* **23**, 246.
Faddeev, L. D., and V. N. Popov: 1973, *Usp. Fiz. Nauk* **111**, 427.
Faddeev, L. D., and A. A. Slavnov: 1978, *Introduction to Quantum Theory in Gauge Fields* (in Russian), Moscow.
Feynman, R. P.: 1963, *Acta Phys. Pol.* **24**, 297.
Frolov, V. P., and G. A. Vilkovisky: 1979, *Preprint IC/79/69*, Moscow; cf. also *Phys. Lett.* **106B**, 307, 1981.
Greenberger, D. M., and A. W. Overhauser: 1980, *Spectrum der Wissenschaft*, July 1980, p. 42.
Grishchuk, L. P., and G. Polnarev: 1980, in: *General Relativity and Gravitation*, A. Held (ed.), Plenum, New York.
Gusev, A. V., and V. N. Rudenko: 1979, *Zhurn. Eksp. Teor. Fiz.* **74**, 819.

Gupta, S.: 1968, *Phys. Rev.* **172**, 1303.
Hawking, S. W.: 1975, *Comm. Math. Phys.* **43**, 199.
Heisenberg, W.: 1930, *Die physikalischen Prinzipien der Quantentheorie*, S. Hirzel, Leipzig.
Heisenberg, W.: 1938, *Z. Phys.* **110**, 251.
Heisenberg, W.: 1967, *Einführung in die einheitliche Theorie der Elementarteilchen*, Hirzel, Stuttgart.
Heisenberg, W.: 1969, *Der Teil und das Ganze*, Piper & Co., München.
Heisenberg, W., and H. Euler: 1936, *Zeitschr. Phys.* **98**, 714.
Heisenberg, W., and W. Pauli: 1928, *Zeitschr. Phys.* **47**, 151.
Heisenberg, W., and W. Pauli, 1929, *Zeitschr. Phys.* **56**, 1.
Heisenberg, W., and W. Pauli: 1930, *Zeitschr. Phys.* **59**, 168.
Hönl, H.: 1981, *Phys. Blätter* **37**, 25.
't Hooft, G.: 1974, *Ann. Inst. Henri Poincaré.*
Infeld, L.: 1953, *Fortschr. Phys.* **1**, 92.
Isaacson, R. A.: 1967, *Phys. Rev.* **166**, 1263, 1272.
Ivanenko, D. D.: 1979a, 'Perennial Actuality of Einstein's Theory of Gravitation', in: H.-J. Treder (ed.), *Einstein-Centenarium*, Akademie-Verlag, Berlin.
Ivanenko, D. D.: 1979b, 'Perennial Modernity of Einstein's Theory of Gravitation', in: F. de Finis (ed.), *Relativity, Quanta and Cosmology in the Scientific Thought of Albert Einstein*, Johnson Reprint Coop., New York, San Francisco, London.
Ivanenko, D. D., and A. Sokolov: 1953, *Klassische Feldtheorie*, Akademie-Verlag, Berlin.
Jordan, P., and W. Pauli, 1928, *Zeitschr. Phys.* **47**, 151.
Källen, G.: 1965, 'Die Entwicklung der neueren relativistischen Quantenfeldtheorie', in: *Physikertagung 1965 in Frankfurt/M.*, B. G. Teubner, Stuttgart.
Kirchhoff, G.: 1859, *Monatsber. Berlin*, p. 662.
Lanczos, C.: 1922, *Phys. Zeitschr.* **23**, 539.
Lanczos, C.: 1924, *Ann. Phys. Leipzig* **74**, 518.
Lanczos, C.: 1938, *Math. Ann. Princeton* **39**, 842.
Lanczos, C.: 1966, 'Tetraden-Formalismus und definite Raum-Zeit-Struktur', in: H.-J. Treder (ed.), *Entstehung, Entwicklung und Perspektiven der Einsteinschen Gravitationstheorie*, Akademie-Verlag Berlin; 1979 in: H.-J. Treder (ed.), *Gravitationstheorie und Theorie der Elementarteilchen*, Akademie-Verlag, Berlin.
Landau, L., and R. Peierls: 1931, *Zeitschr. Phys.* **69**, 56.
Markov, M. A.: 1980, 'On Quantum Violation of Topology in Small Spatial Regions', *Preprint P-0187*, of the Inst. Nucl. Res., Moscow.
Markov, M. A.: 1981, 'On the Maximon and the Concept of Elementary Particle', *Preprint P-0208*, Inst. Nucl. Res., Moscow.
Menger, K.: 1949, 'Die Relativitätstheorie und die Geometrie', in: P. A. Schilpp (ed.), *Albert Einstein: Philosopher — Scientist*, Open Court, La Salle, Illinois.
Mensky, M. B.: 1983, *The Group of Pathes. Measurement, Field, and Particles* (in Russian), Nauka, Moscow.
Mercier, A., H.-J. Treder, and W. Yourgrau: 1979, *On General Relativity*, Akademie-Verlag, Berlin.
Millikan, R. A.: 1916, *Phys. Rev.* **7**, 355.
Misner, C. W., K. S. Thorne, and J. A. Wheeler, 1973, *Gravitation*, W. H. Freeman, San Francisco.
de Muynck, W. M.: 1984, *On the Necessity and Sufficiency of Local Commutativity for Causality in Quantum Mechanics*, Academisch Proefschrift, Amsterdam; *Found. Phys.* **14**, 111, 129 (1984).
v. Neumann, J.: 1955, *Mathematical Foundations of Quantum Mechanics*, Princeton U.P.
Pais, A.: 1982, '*Subtle is the Lord ...*' *The Science and the Life of Albert Einstein.* Oxford U.P.

Papapetrou, A., and H.-J. Treder: 1962, 'Shock Waves in General Relativity', in: *Recent Developments in General Relativity*, Pergamon, Warszawa.
Pauli, W.: 1933, 'Letter to Heisenberg (18 January 1933)', in: K. v. Meyenn (ed.), *Wolfgang Pauli, Scientific Correspondence with Bohr, Einstein, Heisenberg a.o.*, Vol. II (1930–1939), Springer, Berlin, Heidelberg, New York, 1985.
Pauli, W.: 1958, *Theory of Relativity*, Pergamon, Oxford.
Pauli, W.: 1961, *Physik und Erkenntnistheorie*, Vieweg, Braunschweig.
Pechlaner, E., and R. Sexl: 1966, *Comm. Math. Phys.* **2**, 165.
Peres, A., and N. Rosen: 1960, *Phys. Rev.* **118**, 335.
Planck, M.: 1899, *Sitzungsber. Preuss. Akad. Wiss.*, p. 440.
Planck, M.: 1900a, *Verh. Deutsch. Phys. Ges.* **2**.
Planck, M.: 1900b, *Ann. Phys. Leipzig* **1**, 69.
Planck, M.: 1906, *Theorie der Wärmestrahlung*, J. A. Barth, Leipzig.
Planck, M.: 1927, 'Die physikalische Realität der Licht-quanten', in: *Max Planck, Physikalische Abhandlungen und Vorträge*, Vieweg, Braunschweig 1958.
Podolsky, K. B.: 1941, *Phys. Rev.* **62**, 68.
Regge, T.: 1958, *Nuovo Cimento* **7**, 215.
Rosen, N.: 1979, 'Bimetric General Relativity Theory', in: P. G. Bergmann and V. de Sabbata (eds.), *Proceedings of the Intern. School of Cosmology and Gravitation, Erice, May 1979.*
Rosenfeld, L.: 1957, *Report of the Conference on the Role of Gravitation in Physics*, Chapel Hill, North Carolina.
Rosenfeld, L.: 1968, *Proceedings of the 14th Solvay Conference*, p. 232.
Rosenfeld, L.: 1966, 'Quantentheorie und Gravitation', in: H.-J. Treder (ed.), *Entstehung, Entwicklung und Perspektiven der Einsteinschen Gravitationstheorie*, Akademie-Verlag Berlin; 1979, in: H.-J. Treder (ed.), *Gravitationstheorie und Theorie der Elementarteilchen*, Akademie-Verlag Berlin; 1979, in: R. S. Cohen and J. J. Stachel (eds.), *Selected Papers of Léon Rosenfeld*, Boston Studies in the Philosophy of Science, Vol. XXI, D. Reidel, Dordrecht, Boston, London (1979).
Schwinger, I.: 1951, *Phys. Rev.* **82**, 664.
Stelle, K. S.: 1977, *Phys. Rev.* **D16**, 953.
Steenbeck, M., and H.-J. Treder: 1984, *Möglichkeiten der experimentellen Schwerkraftforschung*, Akademie-Verlag, Berlin.
Synge, J. L.: 1960, *Relativity: The General Theory*, North-Holland, Amsterdam.
Tonnelat, M. A.: 1965, *Les Theories Relativistes de l'Electromagnetisme et de la Gravitation*, Michel, Paris.
Treder, H.-J.: 1961, *Monatsber. Akad. Wiss.* DDR **3**, 241.
Treder, H.-J.: 1962, *Gravitative Stoßwellen*, Akademie-Verlag, Berlin.
Treder, H.-J.: 1963, Gravitonen, *Fortschr. Phys.* **11**, 81.
Treder, H.-J.: 1965, 'Die Quantentheorie des Gravitationsfeldes und die Plancksche Elementarlänge', in: *Physikertagung 1965 in Frankfurt/M.*-Höchst, B. G. Teubner Verlagsgesellschaft, Stuttgart.
Treder, H.-J.: 1971, 'The Einstein—Bohr Experiment', in: W. Yourgrau and A. van der Merwe (eds.), *Perspectives in Quantum Theory*, MIT Press, Cambridge, Mass.
Treder, H.-J.: 1974, *Ann. Phys. Leipzig* **31**, 322.
Treder, H.-J.: 1975a, *Ann. Phys. Leipzig* **32**, 383.
Treder, H.-J.: 1975b, *Ann. Phys. Leipzig* **32**, 238.
Treder, H.-J.: 1977, 'Skalen-Invarianz, Eich-Transformation und allgemeine Relativitätstheorie', in: *75 Jahre Quantentheorie*, Akademie-Verlag, Berlin.
Treder, H.-J.: 1979, 'On the Problem of Physical Meaning of Quantization of Gravitational Fields', in: F. de Finis (ed.), *Relativity, Quanta and Cosmology in the Development of the Scientific Thought of Albert Einstein*, Johnson Reprint Coop., New York, San Francisco, London.

Treder, H.-J.: 1982, *Ann. Phys. Leipzig* **39**, 265.

Treder, H.-J.: 1985, *Found. Phys.* **15**, 161.

Treder, H.-J., H.-H. von Borzeszkowski, A. van der Merwe, and W. Yourgrau: 1980, *Fundamental Principles of General Relativity Theories: Local and Global Aspects of Gravitation and Cosmology*, Plenum Press., New York and London.

Unruh, G.: 1978, *Phys. Rev.* **D17**, 1180; **D18**, 1769; 1979, *Phys. Rev.* **D19**, 2888.

Unruh, G.: 1984, in: S. M. Christensen (ed.), *Quantum Theory of Gravitation*, Adam Hilger, Bristol.

Wheeler, J. A.: 1957, *Ann. Phys.* **2**, 604.

Wheeler, J. A.: 1962, *Geometrodynamics*, Academic Press, New York.

Wheeler, J. A.: 1964, 'Geometrodynamics and the Issue of the Final State', in: C. DeWitt and B. DeWitt (eds.), *Relativity, Groups and Topology*. Blackie and Sons, London and Glasgow.

Wheeler, J. A.: 1966, 'Wie steht es heute mit Einstein's Idee, alles als Geometrie zu verstehen', in: H.-J. Treder (ed.), *Entstehung, Entwicklung und Perspektiven der Einsteinschen Gravitationstheorie*, Akademie-Verlag, Berlin.

Wheeler, J. A.: 1968, *Einsteins Vision*, Springer, Berlin, Heidelberg, New York.

Weyl, H.: 1919, *Ann. Phys. Leipzig* **59**, 101.

Weyl, H.: 1923, *Raum, Zeit, Materie*, Springer, Berlin.

Wigner, E. P.: 1982, 'Realität und Quantenmechanik', Lecture read in Lindau and in Tutzing (private communication).

Yang, C. N.: 1974, *Phys. Rev. Lett.* **33**, 445.

Yourgrau, W., H.-H. v. Borzeszkowski, and H.-J. Treder: 1979, *Astron. Nachr.* **300**, 57.

Zel'dovich, Ya. B. and A. A. Starobinsky: 1971, *Zh. Eksp. Teor. Fiz.* **61**, 2161.

Index

GPSR Compliance

The European Union's (EU) General Product Safety Regulation (GPSR) is a set of rules that requires consumer products to be safe and our obligations to ensure this.

If you have any concerns about our products, you can contact us on ProductSafety@springernature.com

In case Publisher is established outside the EU, the EU authorized representative is:

Springer Nature Customer Service Center GmbH
Europaplatz 3
69115 Heidelberg, Germany

Batch number: 10607224

Printed by Printforce, the Netherlands